INTERNATIONAL BENCHMARKING OF
U.S. Chemical Engineering
Research Competitiveness

Panel on Benchmarking the Research
Competitiveness of the US in Chemical Engineering

Board on Chemical Sciences and Technology

Division on Earth and Life Studies

NATIONAL RESEARCH COUNCIL
OF THE NATIONAL ACADEMIES

NATIONAL ACADEMIES PRESS
Washington, D.C.
www.nap.edu

NATIONAL ACADEMIES PRESS   500 Fifth Street, N.W.   Washington, DC 20001

NOTICE: The project that is the subject of this report was approved by the Governing Board of the National Research Council, whose members are drawn from the councils of the National Academy of Sciences, the National Academy of Engineering, and the Institute of Medicine. The members of the committee responsible for the report were chosen for their special competences and with regard for appropriate balance.

This study was supported by the National Science Foundation under Grant CTS-0534814. Any opinions, findings, conclusions, or recommendations expressed in this publication are those of the author(s) and do not necessarily reflect the views of the organizations or agencies that provided support for the project.

International Standard Book Number-13:   978-0-309-10537-8
International Standard Book Number-10:   0-309-10537-4
Library of Congress Control Number 2007927597

Additional copies of this report are available from National Academies Press, 500 Fifth Street, N.W., Lockbox 285, Washington, DC 20055; (800) 624-6242 or (202) 334-3313 (in the Washington metropolitan area); Internet, http://www.nap.edu.

Printed in the United States of America.

PANEL ON BENCHMARKING THE RESEARCH COMPETITIVENESS OF THE US IN CHEMICAL ENGINEERING

Chairperson

GEORGE STEPHANOPOULOS, Massachusetts Institute of Technology

Members

PIERRE AVENAS, ParisTech, Paris, France
WILLIAM F. BANHOLZER, The Dow Chemical Company, Midland, MI
GARY S. CALABRESE, Rohm & Haas Company, Philadelphia, PA
DOUGLAS S. CLARK, University of California, Berkeley,
L. LOUIS HEGEDUS, Arkema Inc. (Retired), Rosemont, PA
ERIC W. KALER, University of Delaware, Newark
JULIO M. OTTINO, Northwestern University, Evanston, IL
NICHOLAS A. PEPPAS, University of Texas, Austin
JOHN D. PERKINS, University of Manchester, United Kingdom
JULIA M. PHILLIPS, Sandia National Laboratory, Albuquerque, NM
ADEL F. SAROFIM, University of Utah, Salt Lake City
JACKIE Y. YING, Institute of Bioengineering and Nanotechnology,
Singapore

National Research Council Staff

TINA M. MASCIANGIOLI, Responsible Staff Officer
ERICKA M. MCGOWAN, Associate Program Officer
JESSICA L. PULLEN, Research Assistant
DAVID RASMUSSEN, Senior Project Assistant
FEDERICO SAN MARTINI, Program Officer
DOROTHY ZOLANDZ, Director

Acknowledgment of Reviewers

This report has been reviewed in draft form by persons chosen for their diverse perspectives and technical expertise in accordance with procedures approved by the National Research Council's Report Review Committee. The purpose of this independent review is to provide candid and critical comments that will assist the institution in making the published report as sound as possible and to ensure that it meets institutional standards of objectivity, evidence, and responsiveness to the study charge. The review comments and draft manuscript remain confidential to protect the integrity of the deliberative process. We wish to thank the following individuals for their review of this report:

Dr. John M. Campbell, Sr., (Retired President and CEO, Campbell Companies), Norman, OK

Dr. Thomas M. Connelly, Jr., E. I. du Pont de Nemours & Company, Wilmington, DE

Dr. Susan Cozzens, Georgia Institute of Technology, Atlanta

Dr. Pablo G. Debenedetti, Princeton University, Princeton, NJ

Dr. Miles P. Drake, Weyerhaeuser Company, Federal Way, WA

Dr. Glenn H. Fredrickson, University of California, Santa Barbara

Dr. Lynn Gladden, University of Cambridge, United Kingdom

Dr. Ignacio E. Grossmann, Carnegie Mellon University, Pittsburgh, PA

Dr. Buddy D. Ratner, University of Washington, Seattle

Dr. James A. Trainham, PPG Industries, Inc., Pittsburgh, PA

Although the reviewers listed above provided many constructive comments and suggestions, they did not see the final draft of the report before its release. The review was overseen by Dr. Maxine Savitz, Retired General Manager of Technology/Partnerships, Honeywell Inc, and Dr. C. Bradley Moore, Northwestern University. Appointed by the National Research Council, they were responsible for making certain that an independent examination of this report was carried out in accordance with institutional procedures and that all review comments were carefully considered. Responsibility for the final content of this report rests entirely with the authors and the institution.

Contents

Executive Summary

This report highlights the main findings of a benchmarking exercise to rate the standing of U.S. chemical engineering relative to other regions or countries, key factors that influence U.S. performance in chemical engineering, and near- and longer-term projections of research leadership. Over a quarter of the jobs in the United States depend on chemistry in one way or another, and over $400 billion worth of products rely on innovations from this field. Chemical engineering, as an academic discipline and profession, has enabled the science of chemistry to achieve this level of significance. However, over the last 10-15 years, concerns have been raised about the identity and future of the U.S. chemical engineering enterprise, stemming from the globalization of the chemical industry; expansion of the field's research scope as it interfaces with other disciplines; and narrowing of the field's ability to address important scientific and technological questions covering the entire spectrum of products and processes—from the macroscopic to molecular level.

At the request of the National Science Foundation, the National Research Council conducted an in-depth benchmarking analysis to gauge the current standing of the U.S. chemical engineering field in the world. The benchmark measures included: (1) the development of a Virtual World Congress comprising the "best of the best" as identified by leading international experts in each subarea; (2) analysis of journals to uncover directions of research and relative levels of research activities; (3) analysis of citations to measure the quality of research and its impact; and (4) the quantitative analysis of trends in degrees conferred to and employment of

1

chemical engineers, and some other measures including patent productivity and awards.

The United States is presently, and is expected to remain, among the world's leaders in all subareas of chemical engineering research, with clear leadership in several subareas. U.S. leadership in some classical and emerging subareas will be strongly challenged. The United States is currently among world leaders in all of the subareas of chemical engineering research identified in the report, and leads in both classical subareas such as transport processes as well as emerging areas such as cellular and metabolic engineering. Although the comparative percentage of U.S. publications has decreased substantially (see Figure 1), the quality and impact still remain very high and clearly in a leading position. For example, 73 of the 100 most-cited papers in chemical engineering literature during the period 2000-2006 came from the United States (see Figure 2). As a result, the United States is expected to maintain its current position at the "Forefront" or "Among World Leaders" in all subareas of

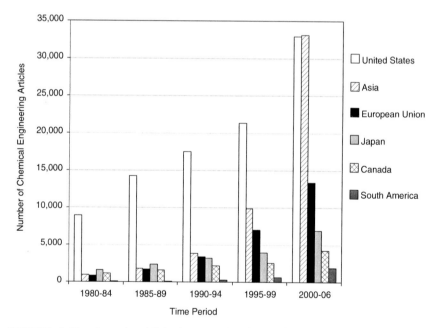

FIGURE 1 Number of published papers in chemical engineering from various geographic regions.
NOTE: Asia comprises China, Korea, Taiwan, and India, and the European Union includes 25 countries.

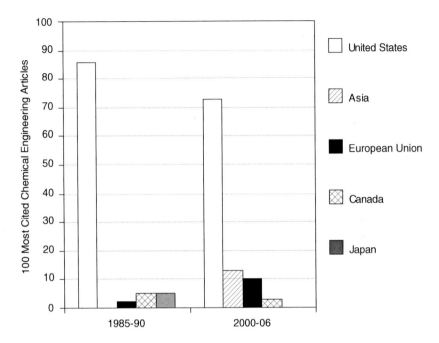

FIGURE 2 Contributions to top 100 mostcited chemical engineering journal articles. In recent decades, the United States has been a strong leader. It is noteworthy that in 2000-2006, 13 of the 100 most cited papers were contributions from Asia. NOTE: Asia comprises China, Korea, Taiwan, and India, and the European Union includes 25 countries.

chemical engineering research, and to expand and extend its current position into subareas such as biocatalysis and protein engineering; cellular and metabolic engineering; systems, computational, and synthetic biology; nanostructured materials; fossil energy and extraction and processing; nonfossil energy; and green engineering.

U.S. leadership in some classical and emerging subareas will be strongly challenged.
      U.S. leadership in the core areas of transport processes; separations; catalysis; kinetics and reaction engineering; process development and design; and dynamics, control, and operational optimization is now shared with Europe and in some cases Japan, as shown by decreases in U.S. journal articles and citations. Japan and other Asian countries are particularly competitive in the materials-oriented research, e.g., polymers, inorganic and ceramic materials, biomaterials, and nanostructured materials. Europe

is also very competitive in the biorelated subareas of research, while Japan is particularly strong in bioprocess engineering. The Panel views the current research trends as healthy. At the same time, it is concerned by the progressive decline of the U.S. position in the core areas, because it is the strength in fundamentals that has enabled generations of chemical engineers to create new and highly competitive technologies for new processes and products.

A strong manufacturing base, culture, and system of innovation, and the excellence and flexibility of the education and research enterprise have been and still are the major determinants of U.S. leadership in chemical engineering.

U.S. chemical, energy, pharmaceutical, biotechnology, biomedical, materials, and electronics companies are well positioned to maintain their effective global presence. Chemical engineering research in the United States is leading to the creation of new technologies and products. Additionally, the chemical engineering education and research enterprise in the United States is excellent, attracting talented people with desired expertise. As shown by the relative fraction of U.S. and non-U.S. publications from chemical engineers, the U.S. also contributes to new areas much faster than other areas in the world, and is better in tune with innovation. At the same time, there is a risk that some of these strengths could erode the traditional core of chemical engineering.

Factors significantly affecting the leadership position of the United States in the future.
The range of chemical engineering research over many spatial and temporal scales, across a broad range of products and processes, and throughout a variety of industries and social needs it serves, has led to innovation and competitiveness but is presently at risk. Most biotechnology and nanotechnology technologies being explored today rely on traditional chemical engineering for implementation. Creating conditions for a more balanced approach that safeguards the dynamic range of chemical engineering research is critical to addressing national needs in energy and the environment and preserving U.S. competitiveness in the future. Future U.S. leadership in chemical engineering is not guaranteed. Many factors could significantly affect the position of the U.S., and these include shifting funding priorities by federal agencies, reductions in industrial support of academic research in the United State, and decreases in talented foreign graduate students, among others.

Summary

According to the American Chemistry Council, over a quarter of the jobs today in the United States depend in one way or another on chemistry, with over $400 billion of products that rely on innovations from this field flowing through the economy.[1] Chemical engineering, as an academic discipline and profession, is a U.S. invention with significant influence from Great Britain that has enabled the science of chemistry to achieve this stunning impact. However, over the last 10-15 years, we have witnessed the following three developments, which have raised many discussions and concerns about the identity and future prospects of the chemical engineering enterprise (education, research, employment):

- drastic restructuring of the global chemical industry and its strategic business philosophy
- continuous expansion of chemical engineering's research scope at the interfaces with several sciences and engineering disciplines such as fluid mechanics, solid particle technologies, polymers, nanostructured materials, protein engineering, biocatalysis, and biomedical devices
- continuous narrowing of chemical engineering's "dynamic range" —or its ability to address important scientific and technological questions covering the entire spectrum from macroscopic to microscopic, to nanoscale, and eventually to molecular scale products and processes, and offer complete solutions

---

[1] See *http://www.americanchemistry.com/s_acc/bin.asp?CID=381&DID=1278&DOC=FILE. PDF.* Accessed February 6, 2007.

The discipline is perceived by its members as being at a crucial time of change, and is presently in the midst of serious and substantive debates on how it should be positioned to meet the needs of the future.

STUDY CHARGE AND PANEL APPROACH

Before addressing questions of whether or not and how chemical engineering should change to meet future needs, it is imperative to understand where the discipline currently is with respect to health and international standing. To that end, a benchmarking exercise was proposed, based on the process established in *Experiments in International Benchmarking of US Research Fields* (COSEPUP, 2000). The discipline has been benchmarked by a Panel of 12 members, 9 from United States and 3 from abroad, with expertise in each of 9 selected areas and an appropriate balance from academia, industry, and national labs. In addition, all the Panel members have extended familiarity of and experience with chemical engineering research not only in Europe but also in Asian countries. Several of the Panel members have setup industrial research centers in Asia (China, India, Japan, Singapore), and all of the Panel members have developed close collaborations with industrial and academic research centers in Europe.

The nine areas of chemical engineering covered in the report are engineering science of physical processes; engineering science of chemical processes; engineering science of biological processes; molecular and interfacial science and engineering; materials; biomedical products and biomaterials; energy; environmental impact and management; and process systems development and engineering. The Panel has considered both quantitative and qualitative measures of the status of the discipline in the above areas and corresponding subareas in response to the following three questions:

• What is the position of U.S. research in chemical engineering relative to that of other regions or countries?
• What key factors influence U.S. performance in chemical engineering research?
• On the basis of current trends in the United States and abroad, what will be the relative future U.S. position in chemical engineering research?

The Panel was asked only to develop findings and conclusions—not recommendations. The Panel focused on leading-edge research, intermixing basic and applied research and process, product, and applications development. The measures used by the Panel include:

- development of a Virtual World Congress comprising the "best of the best" as identified by leading international experts in each subarea;
- analysis of journal publications to uncover directions of research and relative levels of research activities in the United States and the rest of the world;
- comparison of journal submissions by U.S. authors with those by non-U.S. authors;
- analysis of citations to measure the quality of research and its impact;
- patent productivity by academic and industrial research activities;
- analysis of trends in prizes, awards, and other recognitions received by chemical engineers;
- evaluation of leadership determinants such as recruitment of talented individuals to the discipline, funding opportunities, infrastructure, and government-industry-academia partnerships; and
- quantitative analysis of trends in degrees conferred to and employment of chemical engineers.

In an effort to filter out numerical inaccuracies, the Panel opted to rely more on trends than absolute values of these measures. It also based its overall conclusions on the combination of the measures rather than on any single measure. The resulting report details the status of U.S. competitiveness in chemical engineering, by area and subarea. The benchmarking exercise determines the status of the discipline, and extrapolates to determine the future status based on current trends. The Panel does not make judgments about the relative importance of leadership in each area, nor does it make recommendations on actions to be taken to ensure such leadership in the future.

KEY FINDINGS AND CONCLUSIONS

Based on the various benchmarking measures described above, the Panel's principal findings and conclusions can be summarized as follows:

1. The United States is presently, and is expected to remain, among the world's leaders in all subareas of chemical engineering research, with clear leadership in several subareas. U.S. leadership in some classical and emerging subareas will be strongly challenged.

The United States is currently among world leaders in all of the subareas of chemical engineering research, and enjoys a leading position in both classical subareas as well as emerging areas. It is expected that the United States will enhance its relative position in the near future in the following subareas of research:

- biocatlysis and protein engineering;
- cellular and metabolic engineering;
- systems, computational, and synthetic biology;
- nanostructured materials;
- fossil energy extraction and processing;
- non-fossil energy; and
- green engineering.

However, the strong past U.S. position in the follwing subareas, several of which constitute the core of chemical engineering, has been weakened and is expected to continue to weaken in the near future:

- transport processes;
- separations;
- heterogeneous catalysis;
- kinetics and reaction engineering;
- process development and design; and
- dynamics, control, and operational optimization.

Leadership in these subareas is now shared with Europe and in specific instances with Japan, as shown by decreases in journal articles and citations. Japan and other Asian countries are particularly competitive in the materials-oriented research, e.g., polymers, inorganic and ceramic materials, biomaterials, and nanostructured materials. Europe is also very competitive in the biorelated subareas of research, while Japan is particularly strong in bioprocess engineering. The Panel views the current research trends as healthy. At the same time, it is concerned by the progressive erosion of the U.S. position in the core areas, because it is the strength in fundamentals that has enabled generations of chemical engineers to create new and highly competitive technologies for new processes and products.

2. A strong manufacturing base, culture, and system of innovation, and the excellence and flexibility of the education and research enterprise have been and still are the major determinants of U.S. leadership in chemical engineering.

The U.S. chemical, energy, pharmaceutical, biotechnology, biomedical, materials, and electronics companies are well positioned to maintain their effective global presence. This is an essential prerequisite for the continued success of U.S. chemical engineering research. At the same time, chemical engineering research in the United States is creating new platform technologies that may define and propel new classes of products in the market place. This is a relatively new experience for chemical engineering researchers; in the era of commodity chemicals, researchers were concerned with issues of operational efficiency. The U.S. culture and system of innovation are very supportive of these developments.

The chemical engineering education and research enterprise in the United States is excellent. It is diverse, flexible, and agile to competition and attracts talented, young people with experimental, theoretical and computational, academic, industrial, policy-making, and financial and commercial bents. As shown by the relative fraction of U.S. and non-U.S. publications from chemical engineers, the U.S. enterprise also defines or contributes to new areas much faster than its counterparts elsewhere in the world, and is better synchronized with the culture of innovation.

Some of the strengths discussed above are presently at risk with the most important being the risk of progressive erosion in the traditional core of chemical engineering. In this report, the reader will be able to identify the areas at risk, understand why they are at risk, and reach conclusions on what needs to be done.

3. Shifting federal and industry funding priorities, a potentially decreasing ability to attract human talent (domestic or foreign), and a narrowing of the discipline's breadth could diminish the United States' ability to turn today's scientific and technical discoveries into tomorrow's leading jobs in industry and education.

U.S. leadership in the various areas of chemical engineering is not assured for the future. The following factors could have significant effects on the U.S. position:

- shifting funding priorities by federal agencies
- reductions in industrial support of academic research in the United States in favor of academic support in other countries
- potential decreases in the supply of talented foreign graduate students
- reduced attractiveness of chemical engineering as a career path for the most talented U.S. citizens and permanent residents
- shrinking of U.S.-based research laboratories by major chemical companies
- lack of attention to research into methods for shortening the development and implementation cycle for new chemicals, materials, processes, and products

The dynamic range of chemical engineering research over many spatial and temporal scales, across a broad range of products and processes, and throughout the vast variety of industries and social needs it serves, has been a profound force of innovation and competitiveness but is presently at risk. Virtually all of the modern options in biotechnology and nanotechnology being explored today will rely heavily on traditional chemical engineering for implementation. However, if the United States becomes a nation of "nanomaterial-makers," it may be the first to exploit nanomaterials for new energy sources, but the country will lack the wherewithal to implement a total solution. At best, this weakness will only delay implementation; at worst the United States will need to "buy" technology from abroad and suffer the economic consequences. The Panel believes that this issue is of critical importance to addressing national needs in energy and the environment and preserving U.S. competitiveness in chemical engineering in the future.

ASSESSMENT OF CURRENT U.S. POSITION

The current standing of U.S. leadership in chemical engineering research is summarized below in terms of the U.S. position at large and by area of research.

Character of the U.S. Chemical Engineering Research Enterprise (Chapter 2)

Chemical engineering research in the United States covers a spectrum of basic and applied questions, which is far broader than that addressed by chemical engineering researchers in other parts of the world. It is extensively multidisciplinary and interdisciplinary—spanning the conception, design, and development of systems that are primarily based on chemical and biological phenomena. These systems include novel products (chemicals, materials, formulations, and devices) and the processes for making them and using them in various applications. It also includes devising new ways to measure, effectively analyze, and possibly redesign complex systems involving physical, chemical, and biological processes, as in environmental and human health-related research areas.

Chemical engineering research is modestly capital intensive, is deployed through a variety of research modes (e.g., from small single-investigator teams to large multidisciplinary teams), and is supported from a variety of funding sources (e.g., U.S. government, foreign governments, chemical industry,[2] venture capital, and private gifts). Cellular and molecular biology has become an integral core science, and computational approaches are

---

[2]Chemical industry, for the purposes of this report, is the aggregate of companies involved in the production of chemicals, materials, and devices, whose manufacturing or usage involves physical, chemical, or biological processes.

ubiquitous in all areas of research. Chemical engineering research relies substantially on foreign graduate students, who make up from 30% to 70% of the student body at various universities in the United States, and foreign-born research directors at academic institutions and industry.

Assessment of U.S. Position in Chemical Engineering at Large (Chapter 3)

Overall, chemical engineering research in the United States has enjoyed a preeminent position for the last 50 years and is still at the "Forefront" or "Among World Leaders" in every area of research the Panel considered and analyzed. For the last 10 years it has been facing increased competition from the European Union, Japan, and other Asian countries, both in terms of volume of research output as well as quality and impact. Although the fraction of U.S. publications has decreased substantially, the quality and impact still remain very high and clearly in a leading position (e.g., 73 of the 100 most-cited papers in chemical engineering literature during the period 2000-2006 came from the United States). It is anticipated that competition will further increase in the future due to globalization and growth of economies around the world.

Chemical engineering research in the United States is moving away from the traditional core research areas of the discipline and is increasingly focusing on subjects of interdisciplinary interest that interface with applied sciences (physics, chemistry, biology, and mathematics) and other engineering disciplines. Within the scope of these interdisciplinary research activities, it is clearly at the "Forefront," leading the output (volume and quality) of worldwide chemical engineering contributions. However, the fractional volume of output in the core areas of chemical engineering science has been losing ground, and there is serious concern about the discipline's ability to maintain a sufficient number of highly skilled researchers in this area.

Analysis of patents awarded by the U.S. Patent and Trademark Office show that patent productivity of U.S. academic chemical engineering researchers is significantly higher than that of researchers in other countries, and has reached a rough parity with that of U.S. chemistry and materials science and engineering researchers. Also, the relative impact of chemical engineering research on industrial patents has increased.

Assessment of U.S. Position by Area of Research (Chapter 4)

The Panel divided chemical engineering into nine areas of research with three to five subareas in each area. The data indicate that U.S. research is strong and at the "Forefront" or "Among World Leaders" in all subareas of chemical engineering. U.S. research is particularly strong in fundamental engineering science across the spectrum of scale—from macroscopic to molecular. In these areas of research, the primary competition in terms of quality and impact comes from other disciplines rather than chemical engineers from other countries. In the core areas of chemical engineering research, the level of output from European and Asian countries has increased significantly during the last 10 years, but the United States maintains a strong leadership position in terms of quality and impact.

The degree of interdisciplinarity varies from subarea to subarea, but is significant in all areas of chemical engineering research and in recent years has been growing. Therefore, future competitiveness of U.S. chemical engineering research must be benchmarked against a broader spectrum of disciplinary contributions.

KEY DETERMINANTS OF LEADERSHIP

To determine the key factors which influence U.S. performance in chemical engineering research, the Panel collected and analyzed data on recruitment of talented individuals to the discipline, funding opportunities, infrastructure, and government-industry-academia partnerships. The data and their analysis are presented in Chapter 5, and the major findings are summarized as follows:

• Historical research leadership in chemical engineering in the United States is the result of many key factors, the most important of which are excellence and flexibility in education and research; different modes of research, from small, single-investigator teams to large, multidisciplinary teams; a flexible and effective culture and system of innovation; and a strong manufacturing base with global presence.

• Over the years, the United States has been a leader in innovation as a result of a strong U.S. industrial sector, a variety of funding opportunities (industry, federal government, state initiatives, universities, and private foundations), cross-sector collaborations and partnerships, and strong professional societies. Intellectual property policies, administrative support, and access to patent expertise are improving for U.S. academic researchers in chemical engineering. These policies are generally more flexible and advanced here than they are abroad.

• Major centers and facilities have contributed significantly to U.S. leadership by providing key infrastructure and capabilities for conducting research. Key capabilities for chemical engineering research include materials synthesis and characterization, materials micro- and nanofabrication, genetics and proteomics, fossil fuel utilization, and computing facilities.

• There has been an overall steady supply of chemical engineers in the United States, and job prospects and salaries for U.S. chemical engineers are still favorable when compared to those of other sciences and engineering disciplines. However, with changes in U.S. citizenry interests and international capabilities, there is increasingly strong competition for international science and engineering human resources.

- Research funding for U.S. chemical engineering has been rather steady over the years, with an average funding level of approximately $200 million per year between 1993 and 2003. However, during this time the landscape for chemical engineering research has changed significantly and the competitive pressures have increased substantially due to shifting agency priorities.

PREDICTION OF FUTURE U.S. POSITION AND PROJECTION OF KEY DETERMINANTS OF LEADERSHIP

In assessing the future position of U.S. chemical engineering research the Panel took into consideration the following factors (Chapter 4):

- trends in publications and impact, revealed by the analyses in Chapters 3 and 4, which are likely to continue in the near- (2 to 3 years) and mid-term (5 to 7 years) future
- the composition of the Virtual World Congress
- intellectual quality of researchers and ability to attract talented researchers
- maintenance of strong, research-based graduate educational programs
- maintenance of strong technological infrastructure
- cooperation among government, industrial, and academic sectors
- adequate funding of research activities

Prediction of Future Position

U.S. chemical engineering research will remain in the near future strong at the "Forefront" or "Among World Leaders" in all subareas. The Panel foresees that U.S. leadership will be extended in some areas but may be weakened in others. Specifically, the Panel expects that the U.S. position will be strengthened and leadership will increase biocatalysis and protein engineering; cellular and metabolic engineering; systems, computational, and synthetic biology; nanostructured materials; fossil energy extraction and processing; non-fossil energy; and green engineering. The Panel has also recognized that certain developments, for example shifts in government and industry funding priorities and significant investments by European and Asian countries, may put the U.S. leadership position at risk in the following subareas of research: transport processes, separations, catalysis, kinetics and reaction engineering, electrochemical processes, inorganic and ceramic materials, process development and design, and dynamics, control, and operational optimization.

Current government and industry funding priorities will continue to have an impact on chemical engineering's dynamic range, strengthening its molecular orientation in bio-, energy- and materials-related activities at the expense of research in macroscopic processes. Japanese and European research investments maintain a more balanced approach. Also, the growing product- and applications-centric character of the U.S. chemical industry with commensurably increasing levels of applications-oriented research will continue in the future, at the expense of basic research, if no major reorientation of funding priorities by the federal government occurs. Although the United States has enjoyed a research and funding environment that allows for the installation and operation of a diverse range of facilities to support leading-edge research in chemical engineering, this position is not assured forever.

Projection of Key Determinants (Chapter 5)

A steady future supply of highly qualified PhD students in chemical engineering is uncertain.

U.S. chemical engineering departments are still the destination of preference for many foreign graduate students, but as the number and quality of opportunities for research in their home countries continue to improve, the number of talented foreign students coming to the United States may decrease. Also, the number of U.S. citizens and permanent residents pursuing graduate studies in chemical engineering may continue to decrease. Strengthening academic programs and keeping open and exciting research environments that stimulate the intellectual curiosity of young people is essential to attracting and retaining human talent. Salary incentives and more attractive career paths will also be necessary.

The overall federal research and development funding strategy for chemical engineering research is currently unbalanced.

As a result, important developments in key subareas could lag behind in worldwide competition. The dynamic range of the discipline, which has been a principal strength for more than 50 years, is threatened by decreasing support of the traditional core research areas. An important example of this is the inevitable need for alternative energy sources. Virtually all of the options being explored today will rely heavily on traditional chemical engineering for implementation. If we become a nation of "nanomaterialmakers," we may indeed be first to exploit nanomaterials for new energy sources, but we will lack the wherewithal to implement a total solution.

Industry-funded research may have a specific shorter-term focus, and some research projects are conducted under contract terms that capture intellectual properties, protect confidentiality, restrict publication, and require detailed planning and reporting of progress. These conditions may not attract the most talented of the young engineers to the research effort.

Although the United States has enjoyed a research and funding environment that allows for the installation and operation of a diverse range of facilities to support leading-edge research in chemical engineering, this position is not assured forever. Major centers and facilities have contributed significantly to U.S. leadership by providing key infrastructure and capabilities for conducting research. Key capabilities for chemical engineering research include materials synthesis and characterization, materials micro- and nanofabrication, genetics and proteomics, fossil fuel utilization, and cyberinfrastructure. U.S. facilities have instrumentation that is on par with the best in the world. However, rapid advances in design and capabilities of instrumentation can cause obsolescence in 5-8 years. In addition, other countries and regions such as the European Union, Japan, Korea, and China are making heavy capital investments.

1

Background

1.1 CHEMICAL ENGINEERING IN TRANSITION

According to the American Chemistry Council, over a quarter of the jobs today in the United States depend in one way or another on chemistry, with over $400 billion of products that rely on innovations from this field flowing through the economy.[1] Chemical engineering, as an academic discipline and profession, is an American invention that has enabled the science of chemistry to achieve this stunning impact. While George E. Davis of England was the first (in 1880)[2] to publicly discuss the need ". . . to found a distinct branch of the Engineering Profession" that would address the problems facing industry, industrial chemists, and chemical manufacturers at the end of the 19th century, it was in the United States that the nascent concept of chemical engineering was put on firm ground through the groundbreaking introduction of the concept of "unit operations" by Arthur D. Little[3] in 1915:

> Chemical engineering . . . is not a composite of chemistry and mechanical engineering, but a science of itself, the basis of which is those unit

---

[1]See *http://www.americanchemistry.com/s_acc/bin.asp?CID=381&DID=1278&DOC=FILE*. PDF. Accessed February 6, 2007.

[2]J. D. Perkins, "Chemical Engineering—The First 100 Years." Chapter 2 in *Chemical Engineering: Visions of the World*, R. C. Dalton, R. G. H. Price, and D. G. Woods, eds., Elsevier Science B. V., 2003.

[3]Report to the president of the Massachussets Institute of Technology.

operations which in their proper sequence and co-ordination constitute a chemical process as conducted on the industrial scale.

The ensuing development of the first structured educational curriculum at the Massachusetts Institute of Technology (MIT) and the publication of *Principles of Chemical Engineering* (McGraw Hill, 1923), authored by W. H. Walker, W. K. Lewis, and W. H. McAdams, defined the intellectual scope of the new profession and the role of chemical engineers in industry. The MIT Course X was followed quickly by similar educational programs in other universities in the United States and around the world. The subsequent publication of a series of landmark textbooks, *Chemical Process Principles: Part I-Material and Energy Balances* (O. A. Hougen, K. M. Watson, and R. A. Ragatz, 1958), *Mass-Transfer Operations* (R. E. Treybal; 1958), *Transport Phenomena* (R. B. Bird, W. E. Stewart, and E. N. Lightfoot, 1960), *Introduction to the Analysis of Chemical Reactors* (R. Aris, 1965), and others, all originating from U.S. universities, deepened the intellectual scope of the discipline and solidified its American identity. Today, chemical engineers are in central positions determining the course of the chemical industry worldwide.

From its inception, chemical engineering has aimed to respond to and create solutions that satisfy societal needs, as every engineering discipline, almost by definition, does. These societal needs are cumulative; new societal needs arise on top of previous ones. Their evolution over the past 65-70 years, in sequence, includes *defense* (World War II); *living standard and well-being* (creating the petrochemical industry, the "plastics" phenomenon, and scale-up of antibiotics; 1950s); *space and military* (the cold war and accompanying "space race" for satellites, orbiting stations and lunar exploration; 1960s); the *environment* (auto exhaust catalysts, clean air, clean water; 1970s); *energy* (energy crises beginning in the early seventies and reemerging today, alternative forms of energy); *health* (the biotechnology and biomedical revolution; 1970s-1980s); and the *IT revolution* (1990s). These waves have overlapped, creating cumulative effects, have become increasingly globalized, and coupled with technological progress have had the tendency to drive chemical engineering from macroscopic to microscopic, to nanoscale, and eventually to molecular dimensions.

Chemical engineers have been particularly effective at leading these innovations, because they have been trained to think at the molecular level—in terms of chemical, biological, and physical transformations—as well as at the process and system level. As a result, as innovations have moved from macroscopic towards microscopic, and to the nano- and molecular scales, chemical engineering has continued to provide fresh and creative insights and breakthroughs.

Furthermore, the historical dependence of industrial sectors on specific engineering disciplines is changing and chemical engineering has become more important for many industrial sectors than ever before. For example, sustainable energy supply and global warming are acknowledged as key challenges facing the United States and the world. Energy generation was dominated by mechanical and civil engineers (turbine design and project engineering). IGCC (integrated gasification combined cycle) puts a chemical plant at the front of a gas turbine and requires creative solutions from chemical engineering. The medical equipment industry (e.g., computed tomography, magnetic resonance imaging) was dominated by electrical and computer engineering. The next generation of medical diagnostics is "molecular imaging," where we examine the biological/biochemical processes, not just the deformity that occurs as a result of those processes. These are just two of the many examples demonstrating the expanding scope of chemical engineering and its significance in advancing U.S. competitiveness.

Maintaining a *cohesive core* with *intellectual stimulus* has been chemical engineering's most attractive feature for generations of chemical engineers. Its primary strength is the strong basic, yet practical, education it offers, permitting chemical engineers to respond rapidly and in a competent fashion to changing societal and technical demands. The continuous expansion of chemical engineers to an ever-increasing range of scientific and technological problems and their substantive and pivotal contributions to many of them, are testaments to the discipline's powerful intellectual core and its value in addressing a broad range of industrial and societal problems. Chemical engineers are even highly sought after for nontechnical jobs such as investment analysts in the U.S. public equity markets because of their strong ability to think analytically and be effective problem solvers. Educational curricula in nuclear engineering, environmental engineering, biomedical engineering, and biological engineering owe a great deal to chemical engineering's academic core and to chemical engineers who helped their founding.

However, over the last 10-15 years, we have witnessed the following three developments, which have raised many discussions and concerns about the identity and future prospects of the chemical engineering enterprise (education, research, employment):

- drastic restructuring of the global chemical industry and its strategic business philosophy
- continuous expansion of chemical engineering's research scope at the interfaces with several sciences and engineering disciplines
- continuous narrowing of chemical engineering's ability to address important scientific and technological questions across all length scales—its "dynamic range"—as the field has evolved from the macroscopic to the molecular level

Much has been said and written about structural changes in the global chemical industry and how they have affected trends in all aspects of chemical engineering and chemistry. For the purposes of this benchmarking exercise, the following changes are of significance:

- The number of diversified chemical companies has been decreasing, and the ensuing consolidation has led to improvements in operating and financial efficiencies. Commodity chemicals-producing companies faced with large raw materials and energy costs in the United States and slow sales growth are directing their fixed capital investments to regions with large deposits of low-cost raw materials and energy (e.g., the Middle East) or rapidly expanding markets (e.g., Asia, where approximately 50% of the chemicals-consuming markets are expected to be by 2020). Currently there are plans to build about 80 large chemicals plants globally.[4] Each of these plants will require over a billion dollars—and in some cases tens of billions of dollars—to build and none will be built in the United States.

- The R&D outlays for 18 major U.S. chemical companies in 2005 were 2.9% of sales;[5] down from 4.5% in 2000 and more than 5% in the early 1990s. Chemical companies have become more focused and purposeful in their business portfolio and R&D efforts. These developments have had multifaceted effects on chemical engineering research, such as increased efficiency of industrial R&D and new research directions.

- Chemical companies strive for higher value-added products and applications and thus become more sensitive to the market trends. The evolution of the U.S. chemical industry from a process-centered to integrated product-process centered one is of profound significance and has an effect on the type of researchers needed, their educational training, the research directions they pursue, and their career paths.

- Globalization of technology transfer followed globalization of science transfer, which in turn came after globalization of capital flows. Industrial research and development (R&D) centers, under global management, are being established around the world to take advantage of cost-effective human talent that is close to a rapidly growing customer base. Therefore, R&D of new technologies in the chemical industry result from the synergistic efforts of researchers dispersed throughout the world. Local (national) advantages can be derived from low compensation of researchers, high and differentiated talent, academic institutions of world-class quality, availability of venture capital, business-friendly regulations and laws, and

[4] Andrew Liveris, Chairman and CEO, The Dow Chemical Company, address to the Detroit Economic Club, October 30, 2006.
[5] *Chemical Engineering News* 84(6):11-14, February 6, 2006.

progressive general culture, all of which have serious implications on the type of chemical engineering research that a national enterprise follows.

•   The chemical industry in an advanced economy has decidedly changed from a capital-intensive industry to one that relies more and more on knowledge (e.g., scientific, technological, market preferences) as the following developments manifest: (1) industrial R&D outlays are now increasingly considered "investments" and not "expenses," leading the government to redefine how to compute GDP; (2) strong intellectual property (IP) positions now determine the rate of economic success; and (3) IP strategies are today a core part of many corporations R&D strategies.

The intellectual challenge of scientific questions and technological problems at the interface with chemistry, materials science, biology, medicine, electrical engineering, and other disciplines is very attractive and has been drawing chemical engineering researchers in ever-increasing numbers, fueled and supported by accommodating governmental funding policies. In all these areas of interdisciplinary interest, chemical engineers bring a unique combination of analysis and engineering synthesis, which allows them to make contributions with impact far beyond their numbers. It is hard to resist the temptations of these interdisciplinary problems, and it is certainly not advisable to raise obstacles that would discourage them. However, while the *intellectual stimulus* is satisfied by the evolving interdisciplinary research interests, questions about the *cohesion of the disciplinary core* have been raised and need to be answered in a convincing manner. The questions are not addressed by this Panel, but it is clear that they need to be answered to maintain the intellectual cohesion that has propelled chemical engineering research so far. While there is no question that the effort of chemical engineering research at the interfaces with other disciplines has been increasing, there has been no quantitative evidence as to the extent of this shift. The Panel did address this issue.

As chemical engineering research has migrated from the core to the peripheral interdisciplinary research areas, there is a perception that chemical engineering research has been losing its "dynamic range," i.e., its ability to address important scientific and technological questions covering the entire spectrum from macroscopic to microscopic, to nanoscale, and eventually to molecular-scale products and processes and offer complete solutions. As an example, response to the modern energy crisis seems to require more chemical engineers trained in product and process design, electrochemistry, catalysis (heterogeneous), and reaction engineering, all of which are areas that have "peaked" in academic novelty and need to be revitalized in a balanced fashion. Is the perception of decreasing dynamic range in chemical engineering research correct? If yes, to what extent, and what are the consequences on the competitiveness of U.S. chemical engineering research vis-à-vis that of the rest of the world? These are questions that the Panel has asked and explored.

There is a widespread perception that chemical engineering and chemistry are both facing issues of identity and purpose in a time when the disciplines are shifting away from their traditional core and towards areas related to biology, medicine, materials science, and nanotechnology. Concerns about the pipeline of students, the nature of future employment opportunities, and the fundamental health of the disciplines are regular topics of discussion at meetings of the American Institute of Chemical Engineers (AIChE), the American Chemical Society (ACS), and the Council for Chemical Research (CCR), and have been the topic of exercises such as the chemical industry's Vision 2020 and the recent ACS effort, Chemistry 2015. Leaders within the disciplines identify both disciplines as being at a crucial time of change and are struggling with how to position the disciplines to meet the needs of the future. Chemical engineers must also consider the implications for the discipline outlined in the draft NAE report, *Assessing the Capacity of the US Engineering Research Enterprise.*

1.2 STUDY CHARGE AND PANEL APPROACH

Before addressing questions of whether or not and how chemical engineering should change to meet future needs, it is imperative to understand where the discipline currently is with respect to health and international standing. To that end, a benchmarking exercise was proposed, following the process established in *Experiments in International Benchmarking of US Research Fields* (COSEPUP, 2000). The discipline was then benchmarked by a Panel of 12 members, 9 from United States and 3 from abroad, with expertise in each of 9 selected areas and an appropriate balance from academia, industry, and national labs. In addition, all the Panel members have extended familiarity of and experience with chemical engineering research not only in Europe but also in Asian countries. Several of the Panel members have set up industrial research centers in Asia (China, India, Japan, Singapore), and all of the Panel members have developed close collaborations with industrial and academic research centers in Europe.

The nine areas of chemical engineering covered in the report are engineering science of physical processes; engineering science of chemical processes; engineering science of biological processes; molecular and interfacial science and engineering; materials; biomedical products and biomaterials; energy; environmental impact and management; and process systems development and engineering. The Panel considered both quantitative and qualitative measures of the status of the discipline in the above areas and corresponding subareas in response to three questions:

- What is the position of U.S. research in chemical engineering relative to that of other regions or countries?
- What key factors influence U.S. performance in chemical engineering research?
- On the basis of current trends in the United States and abroad, what will be the relative future U.S. position in chemical engineering research?

The Panel was asked to develop only findings and conclusions—not recommendations. They focused on leading-edge research, intermixing basic and applied research and process, product, and applications development. The measures used by the Panel include:

- development of a Virtual World Congress comprising the "best of the best" as identified by leading international experts in each subarea;
- analysis of journal publications to uncover directions of research and relative levels of research activities in the United States and the rest of the world;
- comparison of journal submissions by U.S. authors with those by non-US authors;
- analysis of citations to measure the quality of research and its impact;
- analysis of trends in prizes, awards, and other recognitions received by chemical engineers, chemists, or mechanical engineers;
- evaluation of leadership determinants such as recruitment of talented individuals to the discipline, funding opportunities, infrastructure, and government-industry-academia partnerships;
- quantitative analysis of trends in degrees conferred to and employment of chemical engineers, chemists, or mechanical engineers.

The resulting report details the status of U.S. competitiveness in chemical engineering by area and subarea. The benchmarking exercise determines the status of the discipline, and extrapolates to determine the future status based on current trends. The Panel does not make judgments about the relative importance of leadership in each area, nor does it make recommendations on actions to be taken to ensure such leadership in the future.

In response to the first charge, the Panel assessed current U.S. leadership in chemical engineering research at large and in nine specific areas. The benchmarking results are shown in Chapters 3 and 4, respectively.

The Panel responded to the second question by identifying the determinants of leadership that have influenced U.S. advancement in chemical engineering and the supporting research infrastructure. It also discussed the trends for the future evolution of the key determinants of leadership.

Chapter 5 of the report details the Panel's findings. Chapter 6 provides a summary of the Panel's findings and conclusions. In the final step, the Panel attempted to predict the future U.S. position in chemical engineering at large and in each of the nine specific areas of research. The prediction was based on the assessment of current U.S. positions and trends, as well as the trends in the determinants of leadership and corresponding developments around the world. Chapter 4 of this report includes the Panel's predictions for each of the nine areas assessed.

2

Chemical Engineering Research: Its Key Characteristics, Its Importance for the United States, and the Task of Benchmarking

Engineering is often defined as the discipline that provides a workable and practical heuristic solution to a technical problem within economic, ecological, and time constraints. In this context, the adjective "heuristic" means that the solution is not perfect, because the underlying science is often underdeveloped, but the solution is "good enough" for the purposes intended. For example, before combustion chemistry was understood, well-functioning engines were already made and sold. Because of the heuristic nature of engineering accomplishments, there is always room for technological progress, as science feeds better heuristics.

2.1. WHAT IS CHEMICAL ENGINEERING?

Chemical engineering deals with the engineering aspects of chemical and biological systems of interest. Systems of interest most often include *products, processes for making them,* and *applications for using them.* Beyond designing, manufacturing, and using products, chemical engineering also includes devising new ways to measure, effectively analyze, and possibly redesign complex systems involving chemical and biological processes.

The discipline covers a wide-ranging set of societal interests and needs, including the following: health; habitable environment; national defense and security; transportation; communications; agriculture; clothing and food; and various life amenities. Examples of processes of interest to chemical engineering include a large variety of industrial manufacturing systems used for the production of chemicals and materials (e.g., petrochemical plants, multipurpose pharmaceutical plants, microelectronics fabrication facilities, food processing plants, plants converting biomass to fuels); ecological subsystems such as the atmosphere; the human body in its entirety and its parts; and energy devices such as batteries and fuel cells. Examples of products include various types of commodity or specialty polymers; pharmaceuticals; a broad array of inorganic, ceramic, or composite materials; chemicals and materials for personal care products (e.g., cosmetics, moisturizers, shampoos, antibacterial soaps), information and electronic devices (e.g., displays, cellular phones, optic fiber communication networks), medical products, or automobiles; diagnostic devices; drug delivery systems; and others. Examples of applications include monitoring and control of air pollution; extraction of fossil energy; life-cycle analysis, design, and production of "green" products; diagnostic devices; drug targeting and delivery systems; combustion systems; solar energy; and many others.

Chemical engineering involves the development of heuristic approaches founded on basic science to make it possible to achieve practical outcomes. There is an often discussed overlap between applied sciences (chemistry, biology, and physics) and chemical engineering; often they share the same objective, but use different approaches and methodologies and thus they are synergistic.

Research in chemical engineering seeks to *explain* (analyze) and *control* (synthesize) one or more of the following five basic elements of a system of interest (product, process, or application):

- the physical, chemical, and/or biological phenomena occurring in the system of interest
- the performance of the system of interest, that is, the model-based estimate and/or direct measurement of its properties and usefulness in actual or simulated conditions of application
- the structure and composition of the system of interest that determine the system's properties and performance (e.g., the type of processing units in a manufacturing process and their interconnections, the type of atoms in a chemical product or material and their interconnections, the type of materials and components in a device and their interconnections, the type of reactions in combustion and their interrelationships)
- the synthesis and processing by which a particular product (chemical, material, device) is achieved
- the optimization of any of the above to achieve maximum commercial or societal value

For the purposes of this benchmarking exercise, the Panel divided chemical engineering research into nine major areas with several subareas in each:

Area-1: Engineering Science of Physical Processes
  a. Transport processes
  b. Thermodynamics
  c. Rheology
  d. Separations
  e. Solid particles technology

Area-2: Engineering Science of Chemical Processes
  a. Catalysis
  b. Kinetics and reaction engineering
  c. Polymerization reaction engineering
  d. Electrochemical processes

Area-3: Engineering Science of Biological Processes
  a. Biocatalysis and protein engineering
  b. Cellular and metabolic engineering
  c. Bioprocess engineering
  d. Systems, computational, and synthetic biology

Area-4: Molecular and Interfacial Science and Engineering

Area-5: Materials
  a. Polymers
  b. Inorganic and ceramic materials
  c. Composites
  d. Nanostructured materials

Area-6: Biomedical Products and Biomaterials
  a. Drug targeting and delivery systems
  b. Biomaterials
  c. Materials for cell and tissue engineering

Area-7: Energy
  a. Fossil energy extraction and processing
  b. Fossil fuel utilization
  c. Non-fossil energy

Area-8: Environmental Impact and Management
  a. Air pollution
  b. Water pollution
  c. Aerosol science and engineering
  d. Green engineering

Area-9: Process Systems Development and Engineering
  a. Process development and design
  b. Dynamics, control, and operational optimization
  c. Safety and operability of chemical plants
  d. Computational tools and information technology

It is important to appreciate that the above taxonomy is arbitrary and the various areas and subareas are interrelated and overlapping. For example, combustion research in the subareas "fossil fuel utilization" (Area-7) and "aerosol science and engineering" (Area-8) overlaps extensively with the scope of research in the Area-1 subareas of "transport" and "solid particles technology" and the Area-2 subareas of "catalysis" and "kinetics and reaction engineering." Many materials can be viewed as both "composites" and "biomaterials." The field of complex fluids spans "thermodynamics," "rheology," "molecular and interfacial science and engineering," and subsections of "materials." Loosely speaking, research in Areas- 1, 2, 3, 4, and 9, focuses on fundamentals of engineering science and methodologies, while research in Areas- 5, 6, 7, and 8 focuses on the development of applications (products, processes, devices).

2.1.a What Are the Key Features of Chemical Engineering Research?

Chemical engineering is both multidisciplinary and interdisciplinary.

Almost every process, product, or application that has attracted the discipline's attention involves chemical, physical, and/or biological phenomena at various spatial and temporal scales. Chemical engineering synthesizes knowledge from several disciplines (*multidisciplinary*) and interacts with researchers from multiple disciplines (*interdisciplinary*).

Today in all areas and subareas of interest to chemical engineering, researchers from various disciplines actively compete and collaborate with chemical engineering researchers: applied physicists in fluid mechanics, solid particle technologies, thermodynamics, polymers, rheology, nanostructured materials, protein engineering, molecular and interfacial processes; applied chemists in catalysis, kinetics, all types of materials, molecular and interfacial processes, protein engineering; biologists in biocatalysis, protein engineering, cellular and metabolic engineering, biomaterials, cellular and tissue engineering, biomedical devices, synthetic biology; materials scientists and engineers in all types of materials; and computer scientists, electrical engineers, and operations research and applied mathematicians in all aspects of process systems engineering (modeling, simulation, optimization, control, information technology). All of these scientists and engineers in a scholarly interplay with chemical engineers provide many ideas and motivation for continued growth of components of chemical engineering research.

However, chemical engineering has demonstrated a unique ability to synthesize diverse forms of knowledge from applied sciences and other engineering disciplines into cohesive and effective solutions for many societal needs. This integrative capacity is at the core of the discipline's raison d'etre and is its most distinguishing characteristic.

Chemical engineering research is modestly capital intensive.

The dependence of chemical engineering research on fixed-capital infrastructure varies with area of research. For example, research in all types of materials (Area-5), catalysis (subarea 2a), combustion (subarea 7b), is capital intensive, while research in Area-4 (molecular and interfacial science and engineering) and Area-9 (process systems development and engineering) is not. On average, research in chemical engineering is not as capital intensive as research in materials, but it does involve increasingly sophisticated instruments for the characterization of dynamically evolving reacting systems, chemical and material structures, nano-scale configurations, *in vivo* or *in vitro* characterization of cellular structures and mechanisms, and surfaces and interfaces for a variety of solid and fluid systems. The equipment used in chemical engineering research ranges from small, laboratory bench-scale setups and machines that serve a single investigator to synchrotron sources, nuclear reactors, superconducting magnets, sophisticated surgical facilities, and supercomputers that serve larger user communities and research groups. Chemical engineering research in the United States benefits from the large installed base of research facilities. Europe and Asia have been making significant and sustained fixed-capital investments over the last 10 years. New research centers are being developed with modern facilities, offering chemical engineering researchers in the corresponding regions the necessary infrastructure to compete.

Chemical engineering research is deployed through various modes.

Research problems in chemical engineering require all forms of research, from small-scale research carried out by a principal investigator and a small team, to large multidisciplinary teams and regional consortia involving many investigators. Consortia, alliances, and partnerships of industrial, university, and government laboratories have become fairly common modes in developing and exploiting breakthroughs in the field. Following the globalization of financial markets, globalization of science and technology has increased rapidly and has led to an increasing number of international research collaborations with commensurable sharing of knowledge, fixed-capital, and human resources.

Computational approaches are ubiquitous in chemical engineering research.

Computer-aided research and engineering have been distinctive features of chemical engineering for almost 50 years. Today they are prominent elements of chemical engineering research in all subareas, leading and/or supporting research inquiries from the atomic to the macroscale. The use in chemical engineering research of large supercomputers, networks of computers, sophisticated simulation, control, and optimization packages with ever-improving visualization of the results, and vast arrays of databases, is significant and fast becoming a differentiating strength. Their integration into an effective cyberinfrastructure is the next natural step, and the first attempts in its implementation are currently under way. All areas of chemical engineering research are engaged in simulations of complex phenomena based on first principles at atomic, molecular, meso-, or macroscales, which allow for the prediction of properties and performance and give rise to strategies for the design of processes and materials over the range of relevant scales. In the areas of materials and biotechnology, large databases are mined for the hidden structured knowledge, which can guide the design and control of new materials or cellular and metabolic processes. All of these computer-aided research activities benefit directly from U.S. strengths in computer science and engineering.

Cellular and molecular biology have become core to chemical engineering research.
New discoveries and developments in cellular and molecular biology have led to paradigm shifts in chemical engineering research. New synthetic materials that mimic the structure and properties of naturally occurring ones, new concepts of catalysis using models from protein functions, and new synthetic biological pathways creating new processes are some of the major developments in recent years. Biology has become as core as chemistry and physics have been for the last 100 years of chemical engineering research.

Chemical engineering research requires sustained investment and close interaction with industry.
The time from the first concept to the synthesis of the first prototype to a commercial process, chemical, material, or device, is often as long as, 7 to 15 years. Long-term research is expensive and risky. So, sustained public-sector investment in precompetitive research and development is critical for realizing the economic potential of new ideas. Strong user involvement in the early stages of process or product synthesis and applications-oriented research is pivotal for facilitating the early adoption of a new process, material, chemical, or device.

2.1.b How Important Is It for the United States to Lead in Chemical Engineering and Why?

Chemicals and materials have been central to social advancement and economic growth since the dawn of history. Since World War II there has been an explosion in our understanding of how to make these chemicals and materials, how to use them, and how to adapt them into new products and applications. Chemical engineering and particularly U.S. chemical engineering has been a central force in all of these developments during the past 60 years.

Federally funded research, a strong U.S. chemical industry, and the creative genius of U.S. entrepreneurs catapulted the field into a strong leadership position across the world. Modern refineries; integrated and cost-effective world-class petrochemical processes; chemicals and processes for a continuously advancing agricultural sector; burgeoning pharmaceutical and biotechnology industries; materials, chemicals, and processes for the space program; reductions in air and water pollution; materials and devices that have revolutionized health care practices; computing; telecommunications; and many amenities at home or at the workplace are the historical legacy of U.S. chemical engineers working in collaboration with researchers from other disciplines. It is not an exaggeration to state that almost all aspects of modern life have been impacted by the results of U.S. research in chemical engineering.

The future holds the promise of many exciting dreams: "intelligent" materials that will enable diverse technologies to respond dynamically to changes in the environment; green engineering for a sustainable supply of chemicals, materials, and energy; pharmaceuticals and reconstructive medicine for prolonging human life and improving its quality; intelligent devices for broader and closer interaction among humans worldwide; eradication of many diseases and poverty worldwide; and increased safety and security across the world. In all of these developments, chemical sciences and engineering will continue to play a pivotal role, and thus chemical engineering research will continue to be critical.

To be a leader in industrial growth and to promote a vibrant economy, it is critical that the United States be among the world leaders in all areas of chemical engineering. This requirement implies a dynamic range of chemical engineering research from the molecular to macroscopic scale that has characterized the evolution and past successes of the field. Having world-class researchers who are knowledgeable about the frontiers of chemical sciences and engineering is crucial to the rapid commercial assimilation and exploitation of important discoveries.

Innovations abound in nearly all sectors of our economy, and nearly all modern industries benefit from developments in chemical sciences and engineering research. It is well documented that chemical sciences and engineering together have resulted in the most enabling science/technology combination to underpin technology development in every industrial sector.[1] For example, chemical technology is "Core" in 60% of the 15 broad industrial sectors considered in the study and "Important" in the remaining 40%. It is "Irrelevant" to none of the industrial sectors. No other technology is as prevalent and influential as chemical technology in all industries.

---

[1]Council for Chemical Research, "Measure for Measure: Chemical R&D Powers the US Innovation Engine," 2005.

By comparison, computers and peripherals are "Important" in 8 of the 15 industrial sectors and "Core" in only 4. Additionally, all industries' technologies rely on chemical technology, as is demonstrated by data that indicate that each industry builds on chemical technology as prior art. The evidence is in industry-to-industry patent citation counts; patents granted to companies in all industries build on patents granted to companies in the chemical industry.

Our national defense and security will continue to depend on providing the most advanced diagnostic systems and weapons to our military and police forces. Advanced materials for soldiers' gear, diagnostic devices, portable production or storage of energy, long-range and effective telecommunication devices, and biomaterials and biomedical devices for the wounded, are some of the products to be affected by the results of chemical engineering research to come. Biomaterials are used to make artificial organs, joints, and heart valves, pacemakers, and lens implants, and the range of their applications will continue to grow—impacting treatment processes and delivery of health care in profound ways. Tailored pharmaceuticals and personal care products with minimal side effects, custom design of artificial biological implants that last a lifetime, and processes that make the manufacturing of all of these safe and cost-effective are some of the benefits we can expect. The sustainable supply of chemicals, materials, and energy with minimal impact to the health of the environment and at costs that can be afforded by society is a grand challenge that requires marshalling all of the creative genius of researchers in chemical sciences and engineering.

It is now possible to design new chemicals and materials atom by atom. It is now possible to deliberately and safely engage biological processes to supplement the chain of chemical processes in making the needed materials, chemicals, and devices. The possibilities are seemingly unbounded, but if the United States is to exploit these possibilities, strong national research capabilities by single investigators and multidisciplinary teams are required. Maintaining excellence across the dynamic range of chemical engineering research is essential.

2.2. BENCHMARKING U.S. CHEMICAL ENGINEERING RESEARCH

An engineering research enterprise has multiple objectives. Assessing it is a complex and multifaceted task. Benchmarking it against similar enterprises in other parts of the world is hindered by problems with information sources, which are not necessarily compiled in a comparable manner in other countries and are not readily available in the United States. Valid and useful comparisons are also complicated by the different disciplinary boundaries found in different countries.

The Panel decided that the objectives of the U.S. chemical engineering research enterprise include

•   generation of new fundamental *knowledge* that enlightens the understanding of a broad range of critical engineering problems,
•   generation of new *technologies* that can become the basis for the development of new businesses and enrich the society at large, including not only material goods but also improvements in personal health and the physical environment, and
•   generation of *human resources* with the talents and abilities necessary to meet the challenges of the future.

The following paragraphs in this section will describe the approach the Panel adopted in benchmarking U.S. chemical engineering research against research in other regions of the world. They will also highlight various caveats in benchmarking a research enterprise and how the Panel dealt with these caveats. The results of the benchmarking exercise will be presented in Chapters 3 and 4.

2.2.a Approach

Unlike the basic sciences, whose purpose is to reveal the laws of nature, the purpose of engineering is to provide goods and services for the betterment of life, both individual and collective. Therefore, the objective of chemical engineering research is to create novel, functionally better, or less expensive chemicals, materials, devices, and/or services. Assessing leadership and innovation in chemical engineering research would require measuring value-adjusted rates of (a) creation of new products and services; (b) product, process, and service improvements; and (c) cost reduction through innovation. Such an approach is presently feasible within the confines of a single company but not at an international scale, where detailed data on research and financial performance from a multitude of industrial concerns worldwide are not available.

Therefore, the Panel decided to focus primarily on academic research in chemical engineering, since the results of such an enterprise have a cascading and multiplying effect on (a) the generation of new knowledge underpinning the development of new technologies, (b) the creation of new products and processes, and (c) the formation of human resources that power all of the above. In particular, the Panel selected the following set of metrics to assess the effectiveness of the U.S. chemical engineering research enterprise:

- reputation of U.S. chemical engineering researchers, as manifested by the composition of a Virtual World Congress
  - productivity in publications by U.S. chemical engineering researchers
  - quality and impact of U.S. chemical engineering publications, as measured by the number of citations
  - patent productivity in U.S. chemical engineering departments
  - impact of U.S. chemical engineering research publications in shaping industrial patents
  - impact of U.S. chemical engineering research in developing the requisite high-quality human resources for the advancement of the U.S. chemical industry at large

It should be noted that, in assessing certain subareas, the Panel did take into account the position of U.S. industrial research (see Chapter 4).

As our analysis will show, when all of these metrics are taken together they allow the generation of fairly robust conclusions on the current position and future prospects of the U.S. chemical engineering research enterprise in relation to those in the rest of the world.

Virtual World Congress

A technique used by the Panel to assess leadership in chemical engineering research was to create a Virtual World Congress for each subarea of chemical engineering. Panel members scripted the content of a fictitious World Congress for each subarea of chemical engineering and asked leading experts worldwide to identify 8 to 15 researchers considered to be the "best of the best" in these subareas and likely to make pivotal contributions to the Virtual World Congress. The experts were also asked to develop a short list of "hot topics" in each subarea.

Given the extensive intellectual interaction among the various subareas of chemical engineering and the ensuing cross-pollination, several experts happened to be consulted for the Virtual World Congress in more than one subarea. Furthermore, given the extensive intellectual interaction of almost all chemical engineering subareas with other sciences and engineering disciplines, experts were asked to consider researchers from industry and academia, from other sciences and engineering disciplines.

To ensure that the results of this exercise would reflect chemical engineering research, the Panel decided that at least 50% of the experts selected should be from chemical engineering. No constraints were placed on the fractional representation of chemical engineers in the Virtual World Congress of a specific subarea. A total of 276 individuals participated as experts in this exercise, and the table in Appendix 3A lists their names. The Panel is deeply indebted to them for their effort.

Analysis of Research Publications and Impact

The Panel selected a list of leading journals with significant impact factors for this analysis. Given the broad range of journals in which chemical engineers publish, and in an effort to assess current trends in the direction of chemical engineering research, the Panel selected the journals as follows:

• Journals with broad coverage of chemical engineering research, e.g.,
  o *AIChE Journal*
  o *Industrial and Engineering Chemistry Research*
  o *Chemical Engineering Science*
• Journals with broad coverage of sciences and engineering disciplines in which chemical engineers publish, e.g.,
  o *Science*
  o *Nature*
  o *Proceedings of the National Academy of Science*
• Leading journals for each subarea of chemical engineering:
  o Area-specific journals where researchers from various sciences and/or engineering disciplines publish, along with researchers from chemical engineering, e.g., *Langmuir, Journal of the American Chemical Society, Physics of Fluids*
  o Area-specific journals where chemical engineering researchers are the primary contributors, e.g., *Computers and Chemical Engineering, Journal of Chemical Process Control*

The table in Appendix 3B lists all of the journals considered by the Panel.

The Panel focused its analysis of journal publication data on the following metrics:

• publication rates of growth for the three periods: 1990-1994, 1995-1999, 2000-2006
• percent of U.S. papers in the list of 100 (or 50, or 30, depending on subarea) most-cited papers for the same three periods
• percent contributions by U.S. researchers versus those from other regions
• for subareas of interdisciplinary research, percent contributions by chemical engineers versus those from other disciplines
• for subareas of interdisciplinary research, percent contributions

by U.S. chemical engineers versus those of chemical engineers from other regions

Prize Analysis

The Panel identified the key prizes given in chemical engineering and in various subareas of chemical engineering, and analyzed the list of recipients for each prize. However, it should be noted that most of the prizes have a heavy national or regional bent—very few prizes are truly international. Therefore, the results of this analysis are not truly representative of relative competitiveness of different countries and regions of the world.

Most Significant Advances in Chemical Engineering

The most significant advances in chemical engineering research during the period 1996-2006 were identified, as well as the location where they originated. This information was used by Panel members to assess the relative position of chemical engineering research in the United States in each area and subarea.

All of the above information was used to construct tables that summarize the Panel's assessment, including subjective judgment of the relative significance of numbers, as follows:

(a) What is the current relative position of the United States in each subarea of chemical engineering, using the following scoring system:
"1" Forefront
"3" Among World Leaders
"5" Behind World Leaders

(b) What is the likely future position of the United States in each subarea of chemical engineering, using the following scoring system:
"1" Gaining or Extending
"3" Maintaining
"5" Losing

2.2.b The International Character of Chemical Engineering

To determine the relative competitive strength of U.S. research in chemical engineering, the Panel considered countries—defined by national boundaries—and geographic regions which, due to their specific political or economic links, are clear and distinct competitors. The following countries and geographic regions were considered:

- United States
- Canada
- European Union (of 25 member countries)
- Japan
- Asia (China, Korea, Taiwan, India)
- Central and South America (Mexico, Brazil, Argentina, Chile, Venezuela, Colombia)

In certain instances and in an effort to sharpen the understanding of the competitive landscape, Japan was considered with the other four Asian countries, and China was singled out for comments because of remarkable rates of growth in some areas of Chinese science and engineering. Australia's contributions, significant in certain areas of research, have been considered together with those of the other Asian countries. Contributions from Switzerland, Norway, and Russia were considered as part of the European totals.

The geographic competitive landscape, as described above, is confounded by the rapid advancement of globalization in two ways. First, chemical companies with global reach have established research centers in the United States, Europe, Japan, China, India, Korea, Taiwan, Central and South America, Canada, and Australia. Of particular significance is the recent establishment of many R&D centers in China and India by U.S., European, and Japanese chemical companies. These are not local and self-contained institutions as in the past, but parts of the companies' global R&D organizations. Therefore, R&D of new technologies in the chemical industry result from the synergistic efforts of researchers dispersed throughout the world. Second, the degree of international cooperation in academic research has increased substantially during the last 10 years. For example, the "internationalization index," i.e., the percentage of publications with co-authors from different parts of the world, ranges from 5% to 20% depending on the specific subarea of chemical engineering research (see Chapter 4).

Based on these effects of globalization, the Panel believes that the results of this benchmarking exercise are of value for assessing the future course of U.S. chemical engineering research not only within the confines of the United States but also within the world at large.

2.2.c What Are Some Caveats?

At the outset of this exercise, the Panel recognized a series of caveats in undertaking a project of this scope and magnitude. In the following paragraphs we will lay out these caveats and how the Panel dealt with them. It is important to realize that despite the presence of these caveats, the occasional piece of missing information, and a series of assumptions made by the Panel, the final conclusions possess significant robustness and are supported by several independent lines of analysis with independent sets of data.

Panel Composition

The Panel recognized that its preponderant U.S. constituency (9 of the 12 members are from the United States) might bias its assessment, and it resolved early on to monitor the degree of this bias against other types of independent information. The presence of 3 non-U.S. panel members was very helpful in this regard. In addition, all the Panel members have extended familiarity of and experience with chemical engineering research not only in Europe but also in Asian countries. Several of the Panel members have set up industrial research centers in Asia (China, India, Japan, Singapore), and all of the Panel members have developed close collaborations with industrial and academic research centers in Europe. The Panel believes that its observations and recommendations are quite robust and well founded on the available evidence.

Treatment of Data

In the course of this benchmarking exercise the Panel collected a large amount of numerical data. Most of it is fairly complete, well documented, and indisputable (e.g., data on publications and citations), and only a small part is based on samples of larger data sets (e.g., patent data).

Virtual World Congress

The Panel recognizes that personal biases arising from higher familiarity and interaction with national colleagues could play a role in the recommendations of experts and skew the composition of the Virtual World Congress. Therefore, it has used the relative numbers of participants in conjunction with the numerical results from other sources, e.g., publications and citations. In general, the Panel found that the results of the Virtual World Congress did carry a bias of about 10%-15% but were broadly in line with other indicators.

Publications and Citations

The Panel recognized that analysis of publications by chemical engineers at an international scale is a task complicated by the following two factors:

• The interests of chemical engineering researchers overlap with the interests of researchers from other sciences such as chemistry, biology, applied physics, and applied mathematics, as well as those from other engineering disciplines such as electrical, biological/biomedical/bioengineering, mechanical, civil, or materials sciences. As a result, the Panel opted to refine the search of publications and explore the relative contributions by chemical engineers in the United States and other regions of the world among themselves and vis à vis researchers from other disciplines.

• Affiliation of a researcher with a group that carries the name "chemical engineering" limits the analysis of relative research competitiveness across the world. For example, researchers in certain countries of the European Union and Japan, who are by U.S. definition chemical engineers, are not affiliated with units carrying the name "chemical engineering" in their home institutions. Refinement of the search through the mechanisms available in Web of Science® is difficult, impractical, or impossible. Therefore, the Panel recognizes that a certain ambiguity as to what constitutes a proper comparison of chemical engineering publications by various regions of the world is present throughout Chapters 3 and 4 of this report. To overcome this ambiguity, the Panel has added an analysis of relative competitiveness by U.S. and non-U.S. researchers in each subarea across disciplinary distinctions.

Publication rates and citations per paper vary widely among the various subareas of chemical engineering, and the Panel resisted making broad comparisons of different subareas in terms of these metrics. The only exception is the analysis of publications in the journals with broad coverage of chemical engineering, namely, *AIChE Journal*, *I&EC Research*, and *Chemical Engineering Science*, because the Panel wanted to assess the trends of publication rates in various subareas of chemical engineering over time. The number of papers in the top 100 (or 50, or 30, depending on subarea) most-cited papers in a particular subarea was used as a metric to assess impact. The Panel recognizes the potential pitfalls of such a metric, but it resolved that it is quite representative of relative significance of research contributions, especially if comparisons are limited within the scope of a specific subarea of chemical engineering.

3

Benchmarking Results: Assessment of U.S. Leadership in Chemical Engineering at Large

Chapters 3 and 4 present the results of the benchmarking exercise that the Panel undertook in assessing the international competitiveness of U.S. research in chemical engineering. Chapter 3 summarizes the results for chemical engineering at large, while Chapter 4 presents the results for each subarea of chemical engineering. The approach that the Panel followed for the benchmarking exercise was outlined in Section 2.2.

The presentation of results in this chapter is structured as follows: Section 3.1 describes the composition of the Virtual World Congress (VWC) for each subarea of chemical engineering and draws conclusions on the leadership of U.S. chemical engineering research at large. The detailed analysis of the VWC composition for each subarea is given in Chapter 4. Section 3.2 summarizes the analysis of chemical engineering publications and citations, while Section 3.3 presents the results of a patent analysis. Section 3.4 examines the distribution of prizes, awards, and other recognitions, and Section 3.5 summarizes the Panel's assessment of the current health of U.S. research in chemical engineering at large.

3.1 VIRTUAL WORLD CONGRESS

Table 3.1 summarizes the results of the Virtual World Congress for all subareas of chemical engineering. The table has three parts (from left to right):

(a) The third column of the table presents the total and the relative numbers of U.S. and non-U.S. experts for each subarea.

39

TABLE 3.1 Data for the Virtual World Congress of Chemical Engineering

Area	Subarea
Engineering Science of Physical Processes	Transport processes
	Thermodynamics
	Rheology
	Separation
	Solid particles technology
Engineering Science of Chemical Processes	Catalysis
	Kinetics and reaction eng.
	Polymerization reaction eng.
	Electrochemical processes
Engineering Science of Biological Processes	Biocatalysis and protein eng.
	Cellular and metabolic eng.
	Bioprocess engineering
	Systems, computational, and synthetic biology
Molecular and Interfacial Science and Engineering	Molecular and Supramolecular Assemblies, Micro-Nanopatterned Surfaces and Thin Films
Materials: Molecular Design, Morphology, Processing	Polymers
	Inorganic & ceramic materials
	Composite
	Nanostructured materials
Biomedical Products, Bio-inspired materials, Biomaterials and Biohybrids	Drug targeting and delivery systems
	Biomaterials
	Materials for cell and tissue engineering
Energy	Fossil energy extraction and processing
	Fossil fuel utilization
	Non-fossil energy
Environmental Impact and Management: Safety and Health	Air pollution
	Water pollution
	Green engineering
	Aerosol S&E
Process Systems Development and Engineering	Process development and design
	Dynamics, control, and operational optimization
	Safety and operability of chemical plants
	Computational tools and information technology
TOTAL	

Organizers of Virtual World Congress				Virtual World Congress Speakers (including duplications in nominations)				Virtual World Congress Speakers (excluding duplications in nominations)			
No. of Experts Polled	U.S.	Non-U.S.	% U.S.	No. of Nominations	U.S.	Non-U.S.	% U.S.	No. of Unique Speakers Proposed	U.S.	Non-U.S.	%
7	5	2	71	113	92	21	81	65	50	15	77
11	11	0	100	217	148	69	68	114	70	44	61
8	6	2	75	113	70	43	62	66	34	32	52
9	9	0	100	158	116	42	73	63	41	22	65
6	5	1	83	113	65	48	58	70	36	34	51
7	7	0	100	144	81	63	56	66	33	33	50
8	7	1	88	142	98	44	69	81	51	30	63
11	10	1	91	165	80	85	48	89	39	50	44
5	3	2	60	67	38	29	57	52	26	26	50
7	6	1	86	130	70	60	54	72	30	42	42
8	6	2	75	123	92	31	75	57	43	14	75
7	5	2	71	153	105	48	69	104	65	39	63
9	9	0	100	145	115	30	79	83	65	18	78
15	14	1	93	268	186	82	69	166	105	61	63
20	19	1	95	341	254	87	74	151	102	49	68
14	12	2	86	269	184	85	68	169	106	63	63
10	9	1	90	141	107	34	76	113	79	34	70
14	13	1	93	247	183	64	74	134	88	46	66
11	8	3	73	187	126	61	67	94	60	34	64
10	8	2	80	170	134	36	79	77	62	15	81
6	6	0	100	116	91	25	78	73	55	18	75
4	4	0	100	58	42	16	72	58	42	16	72
8	6	2	75	113	60	53	53	116	65	51	56
6	4	2	67	90	43	47	48	87	43	44	49
7	6	1	86	120	68	52	57	114	65	49	57
7	7	0	100	118	95	23	81	93	78	15	84
10	6	4	60	146	83	63	57	110	74	36	67
5	3	2	60	117	67	50	57	96	57	39	59
16	10	6	63	258	148	110	57	124	73	51	59
12	8	4	67	212	122	90	58	75	49	26	65
11	10	1	91	179	137	42	77	102	70	32	69
7	5	2	71	118	74	44	63	63	41	22	65
296	247	49	83	5051	3374	1677	67	2997	1897	1100	63

(b) The fourth column shows the total number and the percentages of U.S. and non-U.S. participants (speakers) in the VWC for each subarea. These numbers include duplications, i.e., if a specific person was recommended by two experts for the same congress, the entry in the totals is 2.

(c) The fifth column is based on the same information as the middle section, but each participant has been counted once, even if he/she was proposed by several experts.

A total of 296 experts in various areas of chemical engineering were engaged to organize the VWC (see Appendix 3A at the end of this chapter): 83% from the United States and 17% from other countries. For the various subareas the percentage of U.S. organizers ranged from 60% (electrochemical processes; green engineering) to 100% (thermodynamics; separation; catalysis; systems, computational, and synthetic biology; materials for cell and tissue engineering; fossil energy extraction and processing; water pollution), depending on the specific subarea. The preponderance of U.S. names is not surprising given the historical strength of chemical engineering in the United States.

The composition of the resulting Virtual World Congresses, overall and for the various subareas, is the outcome of significance for this benchmarking exercise. As Table 3.1 indicates, 2,997 researchers were recommended for inclusion in the VWC: 1,897 (63%) from the United States and 1,100 (37%) from other countries. The 63% overall U.S. representation in the VWC is in line with the fractional U.S. representation in the list of most-cited publications for 2000-2006 (74%, see Table 3.3), which is a metric that also denotes relative quality and impact. Consequently, the overall composition of the VWC suggests that U.S. chemical engineering research, at large, is "Dominant, at the Forefront" of developments in the discipline.

When we examine the U.S. fractional representation in the VWC for each subarea, we notice that it varies from 42% (biocatalysis and protein engineering) to 77% (transport processes) of all participants, suggesting that U.S. research in every subarea of chemical engineering is either "Dominant, at the Forefront" (65% or more of participants) or "Among World Leaders" (42% to 65% of participants).

In Chapter 4, the specific numbers of Table 3.1 for each subarea are analyzed in conjunction with other information, in order to draw conclusions on the relative position of U.S. chemical engineering research in the corresponding subarea.

3.2 JOURNAL PUBLICATIONS

In this section we will discuss the macroscopic trends, on a worldwide basis, of the publications and citations data collected for five time periods from 1980 to 2006 for the field of chemical engineering at large. Appendix 3B lists all the journals that were considered. They were grouped in the following three categories:

- journals with broad coverage of chemical engineering research
- journals with broad coverage of sciences and engineering disciplines, in which chemical engineers publish
- leading journals for each subarea of chemical engineering

The total number of papers published was found by searching the Web of Science (*http://portal.isiknowledge.com/portal.cgi*) for all publications during the corresponding period, with the requirement that a co-author had a chemical engineering affiliation in the address field. For the United States, a chemical engineering affiliation is a good indicator that a researcher is involved in chemical engineering research. Recent changes in the affiliation of academic researchers from chemical engineering departments to biological engineering or biomedical engineering departments have been taken into account; biological and biomedical engineering departments populated recently by the transfer of chemical engineers were included in the search and the lists were pruned by eliminating the faculty members in these departments who did not hold a Ph.D. in chemical engineering.. However, for non-U.S. researchers with research activities within the scope of chemical engineering as understood in the United States, the corresponding affiliation is not a very good indicator. Many such researchers are affiliated with departments that do not contain "chemical engineering" in their name. Particular attention on select very active universities in Europe and Japan (e.g. ETH-Zurich and Kyoto University, respectively), was given to include the contributions of the non-U.S. researchers who would qualify as chemical engineers, but the numbers of papers by non-U.S. chemical engineering researchers should be viewed as lower bounds.

3.2.a Summary of the Macro Trends

Analysis of publications and citations by chemical engineers in all three groups of journals has revealed the following trends:

- **Trend 1:** The relative volume of the U.S.-originated journal publications by chemical engineers, as a fraction of the worldwide total, has been halved over the past 20 years.

• **Trend 2:** U.S. publications in chemical engineering continue to exercise academic leadership with strong scientific and technological impact worldwide. The relative degree of leadership has been decreasing over the past 10 years.

• **Trend 3:** The relative volume (as a fraction of the total) of U.S.-originated publications in broadly based chemical engineering journals has been reduced by 25%-30% over the past 10-15 years.

• **Trend 4:** The fraction of U.S.-originated contributions, in broadly based chemical engineering journals, with research subjects in the classical coreareas of transport processes, thermodynamics, kinetics and reaction engineering, and process systems engineering, has been reduced by more than the overall fraction in Trend-3, i.e., 50%-60% versus 25%-30% reductions.

• **Trend 5:** The fraction of the top-cited (in the top 100 most-cited papers) U.S.-originated publications in broadly based chemical engineering journals has been reduced by one-third over the past 10-15 years.

• **Trend 6:** The fraction of chemical engineering contributions in broadly based scientific journals, e.g., *Science, Nature, Proceedings of the National Academy of Sciences*, has roughly doubled in the past 5-10 years. Among such contributions U.S.-originated publications represent about 90% of the total.

Taken together, the implications from the above trends are clear:

• **Implication 1:** The volume of research in chemical engineering around the world, especially in the European Union and Asia has been increasing at a higher (European Union) and frantic (Asia) rate compared to that in the United States, but the quality and impact still trail appreciably that in the United States.

• **Implication 2:** Research in U.S. chemical engineering has been driven away from the historical core of chemical engineering toward the periphery, where it meets and overlaps with a variety of other sciences (primarily) and engineering disciplines (secondarily).

• **Implication 3:** While the quality and impact of U.S. research in chemical engineering is still dominant and at the forefront of developments, this leadership position has been weakened over the past 10-15 years, especially in the core areas of the discipline.

In the following paragraphs we will present the details of the data analysis that led to the formation of the above trends and implications.

3.2.b Analysis of Publications and Citations from All Journals

The number of papers from U.S. chemical engineering researchers has dominated the world output over the past 20 years, as Table 3.2 and Figure 3.1 indicate. However, although the number of U.S.-originated publications has increased by a factor of 3.7, its relative contribution to the world total during the past 20 years has been roughly halved from 71% in the period 1980-1984 to 37% in the period 2000-2006 (Trend 1). This is due to a significantly faster growth in the number of publications from chemical engineering researchers across the world. For example, the factor of growth between the 1980-1984 and 2000-2006 periods for various geographic regions is as follows:

- Asia (China, Korea, Taiwan, India): 35
- Central and South America: 23
- European Union (25 countries): 15
- Japan: 4

The 3.7-fold increase in the volume of U.S. publications is primarily the result of an impressive growth in productivity of U.S. researchers, given the fact that the yearly rate of growth in the number of researchers has not increased by a similar factor (see Chapter 5 for trends in numbers of PhD graduates). In contrast, most of the gains in the growth of Asian and European Union publications have come as a result of a significant yearly rate of growth in the number of researchers.

While the relative volume of U.S.-originated chemical engineering publications, as fraction of the world total, has been halved, the academic impact and leadership of the U.S. chemical engineering output has remained at fairly high levels (Trend 2). For example:

TABLE 3.2 Number of Published Papers Originated from Researchers with Chemical Engineering Affiliation at Various Geographic Regions

	1980-1984	1985-1989	1990-1994	1995-1999	2000-2006
United States	8,933	14,230	17,528	21,334	32,899
European Union	890	1,715	3,470	7,015	13,442
Japan	1,647	2,386	3,209	4,022	6,978
Canada	1,182	1,617	2,234	2,605	4,246
South America	81	121	271	651	1,863
Asia (China, Korea, Taiwan, India)	958	1,837	3,907	9,930	33,124

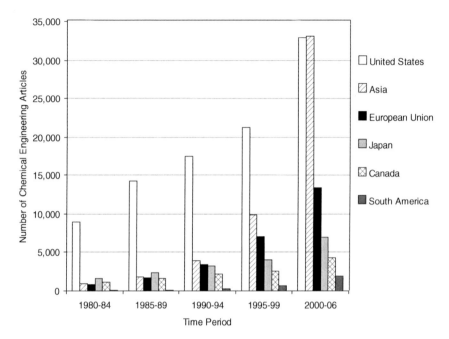

FIGURE 3.1 Number of published papers in chemical engineering from various geographic regions.
NOTE: Asia comprises China, Korea, Taiwan, and India, and the European Union is 25 countries.

- Table 3.3 shows that U.S.-originated publications completely dominated the list of the 100 most-cited publications when the analysis was
carried out for the period 1985-1990: 86 of the top 100, 46 of the top 50,
19 of the top 20, and 10 of the top 10. The analysis for the period 2000-
2006 indicates a continued but weaker dominance of U.S. publications in
the list of the 100 most cited: 73 of the top 100, 37 of the top 50, 13 of
the top 20, and 6 of the top 10. It is worth noting that of the 86 most-cited
U.S. publications with U.S. chemical engineers as co-authors (period, 1985-
1990), 73 had a chemical engineer as the corresponding author, while 13
had a U.S. nonchemical engineer as the corresponding author. In the period
2000-2006, of the 74 most-cited U.S. papers, the corresponding numbers
are 50 with a chemical engineer as the corresponding author and 24 with
a nonchemical engineer as the corresponding author, indicating an appreciable expansion in interdisciplinary research collaboration. This feature
of substantial interdisciplinarity will become more evident later on in this
report in Chapter 4. It is also noteworthy that in the period 2000-2006, no

TABLE 3.3 Most-Cited Papers by Researchers with Chemical Engineering Affiliation (1985-1990 and 2000-2006)

	1985-1990						2000-2006				
	U.S.	EU	Canada	Japan	Australia		U.S.	EU	Canada	Switzerland	Asia
Top 100	86	2	5	5	2	Top 100	73	10	3	1	13
Top 10	10	0	0	0	0	Top 10	6	3		1	
Top 20	19	0	1	0	0	Top 20	13	4		1	2
Top 30	28	0	1	1	0	Top 30	20	4	2	1	3
Top 50	46	1	1	1	1	Top 50	37	6	2	1	4

Japanese contributions were in the top 100 and Asian contributions came from Korea and China.

 • Table 3.4 shows the distribution of the most-cited papers among the various subareas, used to characterize chemical engineering for the purposes of this report, thus underlining the shifts in research emphasis during the past 15-20 years. From the entries of the table it is very clear how the research emphasis has shifted from Area-1 to Areas- 3, 5, and 6. Clearly, numbers of citations vary significantly among the various subareas and may cause uncertainty on the validity of the observed trends. However, these trends will be confirmed with additional data in subsequent paragraphs.

 • It is also interesting to see in what journals the most-cited papers were published. Table 3.5 shows the distribution of the most-cited papers among different groups of papers. These trends will be confirmed with additional data in subsequent paragraphs.

 • The graphs in Figures 3.2 and 3.3 show the evolution of the percentages of published papers from each geographic region with more than 200 and 100 citations, respectively, during the last 20 years. The graph of Figure 3.4, percentage of papers with more than 10 citations, shows a relative parity among the various regions, but this is the group of publications of lesser impact. Note: The numbers in Figures 3.2, 3.3, and 3.4 are percentages of the total number of papers from a given geographic area that satisfy the corresponding citations thresholds.

Clearly, the U.S. dominance in academic impact and leadership, demonstrated by the tables and figures is partly due to historical reasons, that is, to the earlier activity of U.S. researchers compared to that of their Asian and EU counterparts. One would expect that as non-U.S. contributions to archival journals increase, their relative impact will increase as well. Indeed, it noteworthy that of the top 100 most-cited papers, 13 have come from

TABLE 3.4 Distribution of 100 Most-Cited Papers Among the Areas of Chemical Engineering Considered in This Report

Area		Subarea	1985-1990 100 Most-Cited Papers	2000-2006 100 Most-Cited Papers
Engineering Science of Physical Processes	1a	Transport processes	14	2
	1b	Thermodynamics	24	10
	1c	Rheology	7	5
	1d	Separation	10	5
	1e	Solid particles technology	2	0
Engineering Science of Chemical Processes	2a	Catalysis	12	11
	2b	Kinetics and reaction eng.	9	4
	2c	Polymerization reaction eng.	2	6
	2d	Electrochemical processes	0	0
Engineering Science of Biological Processes	3a	Biocatalysis and protein eng.	1	3
	3b	Cellular and metabolic eng.	0	6
	3c	Biochemical engineering	3	0
	3d	Systems, computational, and synthetic biology	0	2
Molecular and Interfacial Science and Engineering	4a		10	12
Materials	5a	Polymers	13	7
	5b	Inorganic and ceramic materials	3	19
	5d	Composite	2	4
	5e	Nanostructured materials	1	11
Biomedical Products and Biomaterials	6a	Drug targeting and delivery systems	3	3
	6b	Biomaterials	1	5
	6c	Materials for cell and tissue engineering	1	7
Energy	7a	Fossil energy extraction and processing	0	0
	7b	Fossil fuel utilization	1	3
	7d	Non-fossil energy	0	1
Environmental Impact and Management	8a	Air pollution	0	0
	8b	Water pollution	1	0
	8c	Aerosol science and technology	0	1
	8d	Green engineering	0	1
Process Systems Development and Engineering	9a	Process development and design	0	0
	9b	Dynamics, control, operational optimization	2	1
	9c	Safety and operability of chemical plants	0	0
	9d	Computational tools and information technology	0	0
TOTAL			122	129
OVERLAP[a]			22	29

[a]The overlap results from accounting the same paper as separate entry in more than one area/subarea.

TABLE 3.5 Distribution of Most-Cited Papers for 1985-1990 and 2000-2006 by Groups of Journals, Indicating Shifts in Direction and Emphasis for Various Subareas of Chemical Engineering

Journals	1985-1990	2000-2006
AIChE J., I&EC Research, Chemical Engineering Science	10	4
Science, Nature, PNAS	9	18
J. Chemical Physics, J. Physical Chemistry, Physical Review Letters, Physical Chemistry-Chemical Physics	13	8
Journal of the American Chemical Society, Accounts of Chem Res.	3	10
Analytical Chem., J. Electron Microscopy, J. Optical Society of America	5	0
Chemical Reviews, Molecular Physics, Fluid Phase Equilibria	5	3
Phys. Reviews Letters, J. Applied Physics, Applied Physics Letters	0	6
J. Catalysis, Advances in Catalysis, Surface Science, Catalysis Reviews, J. of Solid State Chemistry	6	2
Macromolecules, Polymer, J. Polymer Science, Polymer Science and Eng.	12	12
J. Fluid Mechanics, Annual Reviews of Fluid Mechanics, J. Rheology	6	0
Langmuir, J. Colloids and Interfacial Science	4	2
Cancer Research, J. National Cancer Institute	5	0
Biotechnology and Bioengineering	2	0
Nature Biotechnology	0	4
Advanced Materials, Chemistry of Materials	0	5
Biomaterials, Biomacromolecules	0	3

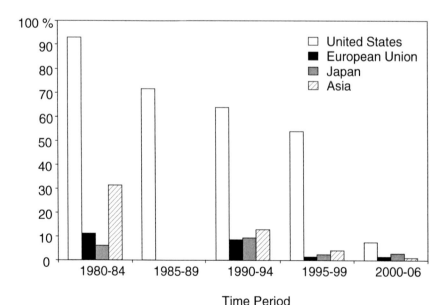

FIGURE 3.2 Percentage of journal articles with 200 or more citations, by region (e.g., 93% of all U.S. publications and 12% of all EU publications during 1980-1984 received more than 200 citations).

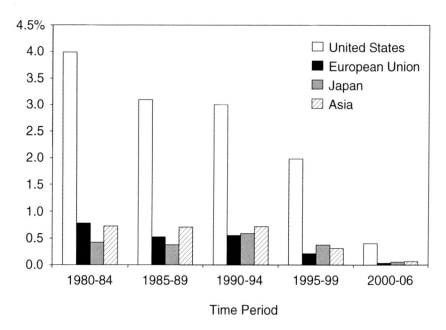

FIGURE 3.3 Percentage of journal articles with 100 or more citations, by region (e.g., 4% of all U.S. and 0.7% of all EU publications during 1980-1984 received more than 100 citations).

the four Asian countries with a relatively brief presence in the international chemical engineering arena.

3.2.c Analysis of Publications and Citations from Journals with Broad Coverage of Chemical Engineering Themes

Three journals have become the main depositories of archival research contributions from a broad spectrum of chemical engineering activities across the world:

- *AIChE Journal*, with a 2005 Impact Factor (IF)[1] = 2.036
- *Chemical Engineering Science*, with an Impact Factor = 1.735
- *Industrial and Engineering Chemistry Research*, with an Impact Factor = 1.504

---

[1]Impact Factor is defined as the number of citations to a journal's published articles during the previous 2 years divided by the number of articles published in the journal.

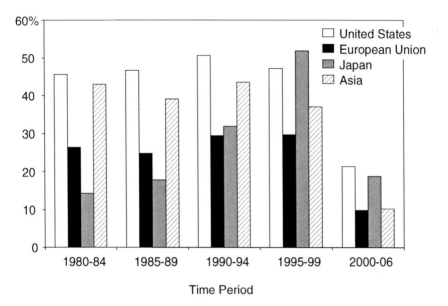

FIGURE 3.4 Percentage of journal articles with 10 or more citations, by region. (e.g., 47% of all U.S. and 43% of all Asian publications during 1980-1984 received more than 10 citations).

Three others, *Chemical Engineering Research and Design* (IF = 0.792), *Chemie Ingenieur Technik* (IF = 0.392), and *Canadian Journal of Chemical Engineering* (IF = 0.574), have a more geographically limited pool of contributions and significantly lesser impact on the leading developments in chemical engineering research. Consequently, our analysis of broadly based trends in research contributions that span the full range of chemical engineering interests were based on data from the first three journals and can be summarized as follows:

• The number of papers contributed from U.S.-based researchers represents a decreasing fraction of all papers published in the three broad chemical engineering journals.
• The fractions of papers contributed from European Union and Asian researchers have been increasing at an appreciable rate.
• Although the U.S. still maintains a very healthy leadership position, the preeminence enjoyed by U.S. contributions during the 1980s (as depicted by the fraction of U.S. papers in the top 100 most-cited papers in each of the three journals) has been eroded.

• Asian contributions are increasing in number and quality, closing the historical gaps quickly.

In the following paragraphs we present the detailed data for the three journals:

• *AIChE Journal*: From Table 3.6 we note that the percentage contribution of published papers originating from researchers in U.S. institutions has been decreasing over the past 20 years, with contributions from the European Union and Asia taking up the difference. However, the U.S. maintains a strong leadership position when one examines the fractions of the top 100 most-cited papers generated by each region (Table 3.7). While the percentage of most-cited papers from the United States has been decreasing, the corresponding percentages of European Union and Asian (China, India, Korea, Taiwan) contributions have been increasing appreciably.

• *Chemical Engineering Science*: Analogous trends are revealed by the analysis of the publications in *Chemical Engineering Science*:
   o The percentage of papers from U.S. authors has decreased over the past 20 years, the contributions from the European Union have

TABLE 3.6 Papers Published in *AIChE Journal*

	1990-1994		1995-1999		2000-2006	
	No. of Papers	%	No. of Papers	%	No. of Papers	%
United States	666	63	786	52	799	43
European Union	132	13	306	20	515	28
Asia	86	8	182	12	312	17
Japan	32	3	68	4	79	4
Canada	59	6	86	6	73	4
South America	10	1	26	2	57	3
Other	67	6	67	4	36	2
Total Papers Published	1,052		1,521		1,871	

TABLE 3.7 Distribution of the 100 Most-Cited Papers in *AIChE Journal*

	1990-1994	1995-1999	2000-2006
United States	78	61	57
European Union	9	22	21
Asia	2	5	9
Japan	4	2	5
Canada	3	6	2

TABLE 3.8 Number of Papers Published in *Chemical Engineering Science*

	1990-1994		1995-1999		2000-2006	
	No. of Papers	%	No. of Papers	%	No. of Papers	%
United States	608	30	570	23	670	19
European Union	703	35	975	40	1,306	38
Asia	237	12	383	16	771	22
Japan	54	3	97	4	183	5
Canada	142	7	157	6	228	7
South America	45	2	71	3	126	4
Other	230	11	207	8	160	5
Total Papers Published	2,019		2,460		3,444	

remained the same level, and the number of papers from Asia has exhibited a marked increase (Table 3.8).

- The U.S. percentage of the 100 most-cited papers (Figure 3.5) has decreased while the European Union percentage has remained at about the same level. However, citations for Asian papers have increased appreciably.

- *I&EC Research*: The data from *I&EC Research* (Table 3.9 and Figure 3.6) reveal a similar picture:

   o The percentage of papers from the United States has been decreasing, while the percentages of contributions from the European Union and Asia have been increasing.
   o The percentages of most-cited papers from the United States has been on the decline, but it still maintains a very healthy leadership position. However, the gap is being closed by an increase in the percentages of most-cited papers from the European Union and Asia.

As the percentage of U.S. contributions to the mainstream chemical engineering journals has been decreasing over the past 20 years, certain core areas of chemical engineering have been affected especially hard. For example, if we define

α number of papers by U.S. authors) / (number of papers by non-U.S. authors)

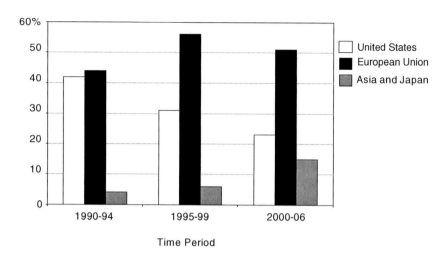

FIGURE 3.5 Distribution of the top 100 most-cited papers in *Chemical Engineering Science*.

TABLE 3.9 Number of Papers Published in *I&EC Research*

	1990-1994		1995-1999		2000-2006	
	No. of Papers	%	No. of Papers	%	No. of Papers	%
United States	860	47	1,139	39	1,705	32
European Union	359	20	707	24	1,448	27
Asia	291	16	497	17	1,183	22
Japan	136	7	168	6	299	6
Canada	93	5	193	7	318	6
South America	38	2	92	3	279	5
Other	49	3	99	4	41	2
Total Papers Published	1,829		2,895		5,273	

then analysis of the data from the three journals indicates that

α (1990-1995; Transport processes) / α (2000-2005; Transport processes) = 3.0

α (1990-1995; Thermodynamics) / α (2000-2005; Thermodynamics) = 3.0

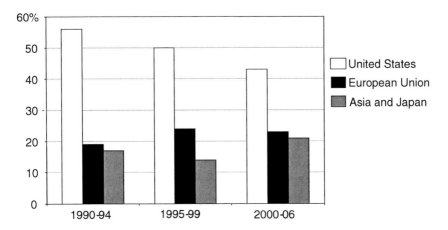

FIGURE 3.6 Distribution of the top 100 most-cited papers in *I&EC Research*.

α (1990-1995; (Process systems eng.) / α (2000-2005; (Process systems eng.) = 2.0

α (1990-1995; Separations) / α (2000-2005; Separations) = 1.5

α (1990-1995; Kinetics reaction eng.) / α (2000-2005; Kinetics reaction eng.) = 2.0

In other words, the representation of U.S. contributions in the above five core areas has been reduced over the past 10 years by a relative factor between 1.5 and 3.0.

3.2.d Analysis of Publications and Citations from Journals with Contributions from Several Disciplines

While the percentage of U.S. research contributions in broadly based chemical engineering journals has been decreasing and European Union and Asian contributions take up a larger share, chemical engineering researchers in the United States have been increasing their presence in scientific journals with contributions from many disciplines, such as *Science, Nature*, and *Proceedings of the National Academy of Sciences*. These are journals with significantly higher impact factors, indicating that U.S. researchers are expanding their reach into areas of science and engineering of increasingly multidisciplinary interest.

In the following paragraphs we present the results of the analysis of data from these three journals:

• SCIENCE: Table 3.10 indicates that contributions from chemical engineers worldwide have increased by a factor of nearly 2, from the 1990-1994 to the 2000-2006 period. A closer look at the number of chemical engineering contributions during the past 5 to 6 years (Figure 3.7) indicates that the doubling of contributions actually occurred during this period. This surge has been led by U.S. chemical engineering researchers, who have contributed 95% of these papers. This is a very strong indicator that the U.S.

TABLE 3.10 Number of Research Papers (Articles) Published in *Science*

	1990-1994		1995-1999		2000-2006	
	No. of Papers	%	No. of Papers	%	No. of Papers	%
Total No. of Papers	4,711		4,985		5,831	
Total No. of Chem. Eng. Papers	51	1	72	1	106	2
U.S., Chem. Eng.	48	95	68	94	101	95
EU, Chem. Eng.	6	11	9	13	19	18
Asia, Chem. Eng.	0	0	3	4	11	10
Canada, Chem. Eng.	0	0	1	1	1	1
S. America, Chem. Eng.	0	0	2	3	2	2
Internationalization (overlap)		5		15		26

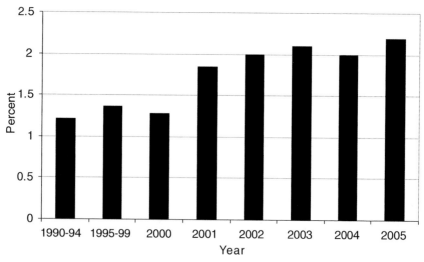

FIGURE 3.7 Percentage of papers in *Science* with chemical engineering co-authorship.

chemical engineering researchers have been leading the exploration of new frontiers and opportunities at the interface with other disciplines. While the percentage of papers that originated from chemical engineering researchers is still low (about 2%), it is nevertheless quite remarkable, given the tremendous competition from other disciplines and the relatively smaller number of chemical engineering researchers.

• *PNAS:* The trends in the *Proceedings of the National Academy of Sciences* publications (Tables 3.11 and 3.12) are similar to those observed in *Science,* namely an increasing fractional representation of chemical engineering contributions, with a dominant percentage of those contributions coming from the United States.

TABLE 3.11 Number of Papers Published in the *Proceedings of the National Academy of Sciences*

	1990-1994		1995-1999		2000-2006	
	No. of Papers	%	No. of Papers	%	No. of Papers	%
Total No. of Papers	11,814		13,053		18,100	
Total No. of Chem. Eng. Papers	46	0.39	82	1	181	1
U.S., Chem. Eng.	46	100	79	96	175	97
EU, Chem. Eng.	1	2	4	5	17	9
Asia, Chem. Eng.	1	2	3	4	12	7
Canada, Chem. Eng.	1	2	2	2	3	2
S. America, Chem. Eng.	0	0	0	0	2	10
Internationalization (overlap)		7		7		15

TABLE 3.12 Number of Papers Published in the *Proceedings of the National Academy of Sciences* during the Past 6 Years

	2000	2001	2002	2003	2004	2005
Total No. of Papers	2,484	2,468	2,680	2,803	2,909	2,847
Total No. of U.S. Papers	1,868	1,872	2,007	2,166	2,170	2,172
% of U.S. Papers	75.20	75.85	74.89	77.27	74.60	76.29
Total No. of Chemical Eng. Papers	11	23	19	29	31	35
% of Chemical Eng. Papers	0.44	0.93	0.71	1.03	1.07	1.23
Total No. of U.S. Chemical Eng. Papers	10	22	19	28	31	34
% of U.S. Papers Among Chemical Eng. Papers	90.91	95.65	100.00	96.55	100.00	97.14

• *Nature:* Chemical engineering contributions during the past 10 years have doubled in number (from 25 in 1990-1994 to nearly 50 in 2000-2006) but their fraction of the total has increased only a little (51% in 1990-1994 to 68% in 2000-2006). More than 90% of the chemical engineering contributions come from U.S. researchers.

3.2.e Analysis of Publications and Citations in Area-Specific Journals

U.S. chemical engineering researchers have been publishing in a long list of journals spanning a very broad range of specific subjects. Depending on the specific area of interest, the contributions of U.S. chemical engineers have varied from 1% (for subareas such as control) to 20% (for subareas such as transport, thermodynamics, catalysis) of the papers published in area-specific journals. Tables 3.13 and 3.14 provide a partial view of the present situation (Appendix 3B lists all the subarea-specific journals whose publications were analyzed). We will not draw any conclusions here from

TABLE 3.13 Chemical Engineering Contributions in Area-Specific Journals (numbers are percentages of papers contributed from chemical engineering researchers in the corresponding journal)

Journal	1990-1994 Total Chem. Eng.	1990-1994 U.S. Chem. Eng.	1995-1999 Total Chem. Eng.	1995-1999 U.S. Chem. Eng.	2000-2006 Total Chem. Eng.	2000-2006 U.S. Chem. Eng.
Fluid Mechanics; Physics of Fluids	7.9	6.9	8.3	7.1	7.8	5.8
Fluid Phase Equilibria; J. Chemical Thermodynamics	27.9	11.2	29.5	9.9	28.8	6.8
Molecular Simulation	14.9	8.5	8.6	6.4	16.4	15.7
J. of Chemical Physics; and J. Physical Chemistry B	7.3	7.0	5.6	5.0	3.8	3.2
J. Catalysis; Applied Catalysis-A and B	21.0	8.4	25.5	12.6	29.5	17.8
Polymer	16.5	5.9	14.5	5.3	10.3	6.2
Progress in Polymer Science	5.5	1.7	3.4	1.3	5.3	0.8
Macromolecules	16.6	10.9	13.5	10.1	11.3	9.3
Metabolic Engineering (2002-06)					36.7	29.1
Enzyme and Microbial Technology	16.0	1.9	15.8	1.8	14.1	6.2
AUTOMATICA	3.5	1.7	4.1	1.9		
IEEE Transaction on Automatic Control	0.8	0.5	1.2	1.0	1.2	1

TABLE 3.14 Chemical Engineering Contributions in Area-Specific
Journals (partial list)

	1995-2006	
Journal	Chem. Eng. (% of published papers)	U.S. Chem. Eng. (% of published papers)
J. of Rheology	30-50	15-20
J. Non-Newtonian Fluid Mechanics	20-30	9-12
Rheological Acta	25-35	10-12
International Journal of Multiphase Flow	10-12	3-5
Powder Technology	25-30	8-10
Catalysis Today	15-25	3-5
J. Applied Electrochemistry	15-20	2-3
Proteins: Structure, Function and Bioinformatics	1-2	0.90-1.4
Protein Science	2-4	~ (2-4)
Langmuir	15-20	5-10
J. Colloid and Interface Science	15-20	10-15
Advanced Materials	5-15	5-10
Chemistry of Materials	5-10	6-8
J. American Ceramic Society	2-6	2-5
Polymer Composites	10-20	5-10
Nano Letters	6-8	~ (6-8)
J. Controlled Release	8-10	7.5-9.5
Advanced Drug Delivery Reviews	3-5	~ (3-5)
Pharmaceutical Research	2-5	~ (2-5)
J. Biomaterials Science: Polymer Edition	15-25	8-12
Biomacromolecules	8-10	7-9
J. Biomedical Materials Research	7-10	5-9
Lab-on-a-Chip	7-9	6.5-8.5
Combustion and Flame	5-10	2-3
Fuel	20-25	3-5
Combustion Science and Technology	3-8	2-4
Combustion Theory and Modeling	4-8	2-4
Environmental Science and Technology	5-8	3-4
J. Environmental Engineering	4-10	2-6
Aerosol Science and Technology	10-20	7-15
J. Aerosol Science	5-12	2-6
SIAM Journal on Optimization	~ 1	~ 1
SIAM Journal on Scientific Computing	1-2	1-2
Optimization and Engineering	5-15	4.8-14.5
Mathematical Programming	0.8-1.0	0.9-1.0
J. Optimization Theory and Applications	0.5	0.5
J. Global Optimization	3.5-4.0	3.4-3.9
INFORMS J. on Computing	0.5	0.5
Annals of Operations Research	1	1

these data, because the sections in Chapter 4 will provide a more detailed analysis of the publication trends in specific subareas and will describe the relative position of U.S. contributions versus those of other geographic regions.

3.3 PATENT PUBLICATION ANALYSIS

Chemical sciences and engineering have resulted in the most enabling science/technology combination to underpin technology development in every industrial sector, as a study sponsored by the Council for Chemical Research (CCR) has revealed.[2] Indeed, as the U.S. Patent and Trademark Office data in the CCR report indicate, each industry builds on chemical technology as prior art. Furthermore, the CCR study has found that chemical companies with highly cited patents have stronger financial performance than companies with lower impact patents; their stock prices, operating revenues, and profits are 35%-60% higher, on average. Additionally, companies that invest in high-quality technology that continues to influence the technological directions of the chemical industry have the most favorable financial performance.

All of the above observations have a direct linkage with the capacity of the U.S. chemical engineering research enterprise to deliver scientific results for high-quality patents and produce first-rate human resources. In this section we will examine the competitiveness of U.S. chemical engineering research in producing technological knowledge for patents with high impact.

Clearly, a complete and authoritative study linking U.S. research in chemical engineering to high-impact patents, where impact is measured by the financial performance of the chemical companies driven by these patents, is an overwhelming task and beyond the charge of this panel. Most of the necessary information for such study cannot be disaggregated from financial results, which in their raw form are not available to the general public. Therefore, the Panel opted to generate indirect evidence by asking the following questions:

• What is the productivity of U.S. chemical engineering departments in generating patents and how does it compare to the productivity of non-U.S. research institutions?

• What is the impact of U.S. academic chemical engineering research in the formation of industrial patents? How does it compare to the impact of non-U.S. chemical engineering research?

---

[2]Council for Chemical Research. "Measure for measure: Chemical R&D powers the US innovation engine," 2005.

- What is the competitiveness of the U.S. industrial research in generating high-impact patents?

By themselves, the answers to these questions are of little significance, but when taken together with the other metrics used in this study, e.g., publications, citations, and the VWC, they can contribute to a better-rounded overall assessment of the U.S. competitiveness in chemical engineering research.

3.3.a Patent Productivity of U.S. Chemical Engineering Departments

In 2003 approximately twice as many patents (3,259) in all fields were awarded to all U.S. universities and colleges as in 1994 (1,783).[3] The total for the 10-year period is 27,594 (16,545 to public and 10,321 to private institutions). Similar trends have been observed in chemical engineering.

Data from five U.S. chemical engineering departments, with significant numbers of patents awarded, indicate that they produced from 2 to 7.5 patents per active research faculty over a period of 20 years, leading to an index of 0.1 to 0.38 patents per active research faculty per year. These numbers indicate that the following:

- Patent productivity of active chemical engineering departments is comparable to that in chemistry (0.25 patents per faculty per year; data from three high-ranking chemistry departments) and materials science and engineering (0.35 patents per faculty per year; data from three high-ranking departments).
- Before 1995-2000, patent productivity of non-U.S. chemical engineering departments had been very low, due to a lack of well-organized Technology Licensing Offices (TLO) within non-U.S. universities. Since 2000 the number of non-U.S. universities with well-organized and purposeful TLOs has increased substantially, especially in Japan and Western Europe. A sample of two European Union and two Japanese chemical engineering departments, all with excellent reputations for academic research, indicated that the corresponding index values are 0.05 to 0.1 patents per faculty per year, significantly lower than that of the active U.S. departments.

3.3.b Impact of Academic U.S. Chemical Engineering Research on Industrial Patents

To study the impact of academic chemical engineering research on industrial patents, the Panel collected the following data:

---

[3] See *http://www.nsf.gov/statistics/seind06/append/c5/at05-68.xls.*

- references of prior art in all of the patents of two major U.S. chemical companies for the years 1995, 2000, and 2005
- references of prior art in a sample of 500 patents from five chemical companies (three from the United States, one from Europe and one from Japan) awarded by the U.S. Patent Office during the period 1995-2003

The set of seven companies represented commodity production and specialty chemicals and materials production. No pharmaceutical companies were included in the set. The groups of patents examined covered both material structure and process patents.

The results from the analysis of the data are as follows:

- The percentage of patents with at least one reference to a publication in a scientific journal (i.e., an indication of linkage to academic research) varied from 12% to 60%, with the higher percentage indicating a patent with higher scientific linkage.
- The percentage of references to scientific journals over the total number of references to prior art varied from 12% to 20%.
- The percentage of references to published chemical engineering papers over the total number of scientific references varied from 4% to 11%.
- The percentage of references to U.S.-originated chemical engineering publications over the total number of references to all chemical engineering publications varied from 45% to 70%.

The limited size of the analyzed set of patents notwithstanding, the above numbers suggest the following conclusions:

- Publications of academic chemical engineering research appear with a frequency of 1 in 9 to 1 in 25 scientific references of the industrial patents examined.
- The dominance of U.S. chemical engineering publications, among all chemical engineering publications, in shaping industrial patents is quite clear; 1 in 2 to 2 in 3 references are for U.S.-originated publications. This is in agreement with the strong presence of U.S. publications in the lists of the most-cited papers, discussed in Section 3.2.

We should clearly recognize that the above analysis has been based on sets of patents awarded by the U.S. Patent Office and as such the results will undoubtedly be somewhat biased. Patents issued by, for example, the Japanese Patent Office may show a different picture. However, most of the high-impact patents filed and awarded by patent offices around the world are also filed in the United States through patent, trademark, and copyright mechanisms. Therefore, the above results are quite credible as representative of existing trends.

3.3.c Competitiveness of U.S. Industrial Research in Generating High-Impact Patents

It is not the purpose of this paragraph to provide a detailed account of the U.S. competitive patent position across the various segments of the chemical industry. Instead, the panel wanted to examine whether the significant strength of the U.S. academic chemical engineering research enterprise (e.g., composition of the VWC, publications, and citations) was reflected in the strength of U.S. industrial research, manifested by strong and differentiating intellectual property position.

First, we note that data on patents in the CCR study, "Measure for Measure: Chemical R&D Powers the U.S. Innovation Engine," indicate that the impact of U.S.-invented chemical technology patenting (number of citations) has risen steadily, in contrast to the declining impact of Japanese-invented chemical technology patents and the steady but relatively lower impact of German-invented chemical patents.

Second, data were collected for the geographic origins of inventors of patents awarded by the U.S. Patent Office during the period 1985-2005 for a few selected areas of chemical engineering related technological research. The percentages of U.S.-invented patents are shown in Table 3.15.

From the table below it is clear that U.S. companies are in a leadership position in generating intellectual property, but it is also clear that there has been a worrisome decline in the U.S. percentage of generated patents over the past 20 years. U.S. chemical companies have recognized this development and are stepping up their efforts in intellectual property generation.

Furthermore, Table 3.16 summarizes the percentages of U.S. patents awarded to U.S., European, and Asian companies in the areas of industrial

TABLE 3.15 Percentages of U.S.-Invented Patents, Awarded by the U.S. Patent Office, in Various Areas of Research

Area of Research	1985-1989	1990-1994	1995-1999	2000-2005
Heterogeneous Catalysts	80	60	45	40
Homogeneous Catalysts	55	60	50	45
Polymerization	55	50	41	51
Fluid Flow Related	80	62	46	42
Fermentations	60	52	55	53

TABLE 3.16 Percentages of Patents, Awarded by the U.S. Patent Office, to U.S., EU, and Asian Assignees in Three Areas of Chemical Industry for the Years 1995, 2000, and 2005

Area of Technology	1995			2000			2004		
	U.S.	EU	Asia	U.S.	EU	Asia	U.S.	EU	Asia
Industrial Separations	55	31	14	53	29	18	56	26	19
Composites	60	13	27	53	18	30	54	18	28
Ceramics	51	22	27	48	22	30	49	18	33

separations, composites, and ceramic materials. These data show a strong U.S. intellectual property position in all three areas of technology.

3.4 PRIZES, AWARDS, AND RECOGNITIONS

There are no international prizes or awards, which recognize research contributions in chemical engineering at large, akin to the Nobel Prize in Chemistry. Therefore, we cannot use this metric for a direct comparison of U.S. versus non-U.S. contributions in chemical engineering.

However, there is a series of international awards and prizes recognizing research contributions in specific subareas of chemical engineering. In addition, there are national awards open to foreign contestants from several disciplines. The Panel has collected data on the winners for a number of such awards in an effort to assess U.S. leadership in specific subareas of chemical engineering research across disciplines and across geographic regions.

The data suggest the following conclusions:

• U.S. chemical engineering researchers have received a significant number of prestigious awards with international competition over a broad range of research subareas; fluid mechanics, catalysis, controlled drug release, bioprocesses, aerosol science and engineering, rheology, reaction engineering, combustion, and materials. These awards seem to confirm earlier observations that breadth and depth (quality) co-exist in U.S. chemical engineering research.

• U.S. chemical engineering researchers have been very competitive with researchers from other disciplines, drawing a significant number of U.S. awards in all subareas of chemical engineering from various disciplinary organizations. Again this information confirms earlier observations on the interdisciplinary competitiveness of U.S. chemical engineering researchers.

3.5 SUMMARY

The analyses in the previous sections has largely been confined to macroscopic trends and assessment of U.S. research in chemical engineering at large. In Chapter 4 each subarea of chemical engineering research is assessed separately, using the same metrics.

Using the results of the overall assessment, discussed in earlier sections of this chapter, as well as the summary overview for all subareas, given in Table 4.45 and discussed in detail in Chapter 4, we can draw the following conclusions regarding the state of U.S. chemical engineering research at large:

- **Conclusion 1:** It has enjoyed a preeminent position for the past 50 years and is still at the "Forefront" or "Among the World Leaders" in every subarea of chemical engineering research.
- **Conclusion 2:** For the last 10 years it has been facing increased competition from European Union and Asian countries, both in terms of volume of research output as well as quality and impact. Although the percentage of U.S. publications has decreased substantially, the quality and impact still remain very high. It is anticipated that competition will further increase in the future due to globalization and growth of competing economies.
- **Conclusion 3:** It has been losing ground in the core areas of chemical engineering (transport processes, thermodynamics, kinetics and reaction engineering, and process systems engineering), which raises concern for its capacity to maintain a sufficient number of highly skilled researchers in these areas.
- **Conclusion 4:** It has been moving away from the core research areas of the discipline and is increasingly focusing its attention on subjects of interdisciplinary interest at the interface with applied sciences (physics, chemistry, biology, mathematics) and other engineering disciplines. Within the scope of these interdisciplinary research activities, it is clearly at the "Forefront," leading the output (volume and quality) of worldwide chemical engineering research.
- **Conclusion 5:** It has been generating an increasing number of patents with continuously increasing commercial impact. Patent productivity of U.S. academic chemical engineering researchers is significantly higher than that of researchers in other countries, and has reached rough parity with that of U.S. chemistry and materials science and engineering. Also, its relative impact on industrial patents has increased.

APPENDIX 3A

Experts Who Organized the Virtual World Congress by Nominating Its Keynote Speakers

EXPERT (VWC Organizer)	AFFILIATION
Agassant, Jean-Francois	ENSMP (France)
Agrawal, Rakesh	Purdue University
Aizenberg, Joanna	Alcatel-Lucent Technologies
Allen, David	University of Texas-Austin
Anseth, Kristi	University of Colorado
Arastopour, Hamid	Illinois Institute of Technology
Arendt, Steve	ABS Consulting
Arkin, Adam	LBNL/UC Berkeley
Arnold, Frances	California Institute of Technology
Athanassiou, Kyriacos	Rice University
Avidan, Amos	Bechtel, USA
Azapagic, Adisa	University of Surrey (UK)
Baer, Eric	Case Western Reserve University
Bakshi, Bhavik	Ohio State University
Barteau, Mark	University of Delaware
Basaran, Osman	Purdue University
Bashir, Rashid	Purdue University
Bates, Frank	University of Minnesota
Baxter, Larry	Brigham Young University
Beer, Janos	MIT
Bell, Alexis T.	UC Berkeley
Berger, Scott	AIChE
Betenbaugh, Michael	Johns Hopkins University
Bizios, Rena	University of Texas-San Antonio
Blanch, Harvey	UC Berkeley
Blankschtein, Daniel	MIT
Blau, Gary	Purdue University
Blum, Frank	University of Missouri-Rolla
Bonvin, Dominique	EPF Lausanne
Bowman, Chris	University of Colorado
Brannon-Peppas, Lisa	University of Texas-Austin
Brinker, Jeffrey	Sandia National Labs
Buttrey, Douglas	University of Delaware
Cairns, Elton	LBNL/UC Berkeley
Caram, Hugo	Lehigh University
Carberry, John	Dupont
Chakraborty, Arup	MIT
Chen, Bingzhen	Tsinghua University
Chmelka, Bradley	UC Santa Barbara
Chornet, Esteban	Usherbrooke (Canada)
Chum, Stepen	Dow Chemical
Clift, Roland	Surrey University, UK

EXPERT (VWC Organizer)	AFFILIATION
Coates, Geoff	Cornell University
Cohen, Yoram	UCLA
Cooper, Stuart	Ohio State University
Coppens, Marc-Olivier	Technische Universiteit Delft
Corn, John	Ohio State University
Couvreur, Patrick	University of Paris
Crowl, Dan	Michigan Technological University
Dal Pont, Jean-Pierre	ESPCI (France)
D'Alessio, Antonio	University of Naples
Dam-Johansen, Kim	DTU (Denmark)
Davis, Mark	California Institute of Technology
Dealy, John	McGill University (Canada)
Debenedetti, Pablo	Princeton University
Denn, Morton	City College of New York
dePablo, Juan	University of Wisconsin
deSmedt, Stefaan	University of Ghent
DiSalvo, Frank	Cornell University
Dixit, Ravi	Engineering and Process Sciences
Doherty, Michael	UC Santa Barbara
Dordick, Jonathan	RPI
Drzal, Lawrence	Michigan State University
Dudukovic, Michael	Washington University
Dumesic, James	University of Wisconsin
Eckert, Charles	Georgia Tech
Edgar, Thomas	University of Texas-Austin
Edwards, David	Harvard University
Eldridge, Bruce	University of Texas-Austin
Fan, L.S.	Ohio State University
Feinberg, Martin	Ohio State University
Floudas, Christodoulos	Princeton University
Flytzani-Stephanopoulos, Miretta	Tufts University
Forrest, Stephen	University of Michigan
Francis, Lorraine	University of Minnesota
Frank, Timothy	Dow Chemical
Fredrickson, Glenn	UC Santa Barbara
Friedlander, Sheldon K.	UCLA
Froment, Gilbert	Texas A&M
Fuller, Gerry	Stanford University
Gani, Rafique	Technical University of Denmark
Gasteiger, Hubert	University Duesseldorf
Genzer, Jan	North Carolina State University
Georgiou, George	University of Texas-Austin
Gandhi, Harendra	Ford Motor Co
Glaborg, Peter	DTU (Denmark)
Gladden, Lynn	Cambridge University
Goodenough, John	University of Texas-Austin
Gooding, Charles	Clemson University

EXPERT (VWC Organizer)	AFFILIATION
Gorte, Raymond	University of Pennsylvania
Gottesfeld, Shimshon	MTI MicroFuel Cells Inc.
Graham, Mike	University of Wisconsin
Green, Don	University of Kansas
Grossmann, Ignacio	Carnegie Mellon University
Gschwend, Philip M.	MIT
Gubbins, Keith	North Carolina State University
Hall, Carol	North Carolina State University
Haller, Gary	Yale University
Hammond, Paula	MIT
Hangleiter, Andreas	Technische Universitat Braunschweig
Harold, Michael	University of Houston
Hawker, Craig	UC Santa Barbara
Haynes, Brian	University of Sydney (Australia)
Haynes, Charles	University of British Columbia (Canada)
Hendershot, Dennis	Chilworth Technology
Heuer, Arthur	Case Western Reserve University
Hidy, George	Envair/Aerochem
Hill, Michael	University of Massachusetts
Hilt, J. Zach	University of Kentucky
Hines, Melissa	Cornell University
Hoo, Karlene	Texas Tech University
Howard, Jack B.	MIT
Hubbell, Jeffrey	Ecole Polytech Fed Lausanne
Iglesia, Enrique	UC Berkeley
Israelachvili, Jacob	UC Santa Barbara
Jachuck, Roshan	Clarkson University
Jain, Pradeep	University of Florida
Jimenez, Jose Luis	University of Colorado
Johansen, Kim Dam	Technical University of Denmark
Johnston, Keith	University of Texas-Austin
Jorne, Jacob	University of Rochester
Kauppinen, Esko I.	Helsinki University of Technology (Finland)
Keasling, Jay	UC Berkeley
Khakar, Devang Vipin	Indian Institute of Technology
Khan, Saad	North Carolina State University
Kletz, Trevor	Loughborough University (UK)
Klibanov, Alexander	MIT
Klimov, Victor	LANL
Kohlbrand, Henry	Dow Chemical
Konstantinov, Konstantin	Bayer Corp
Kopecek, Jindrich	University of Utah
Krishnamoorti, Ramanan	University of Houston
Ladisch, Mike	Purdue University
Lahti, Paul	University of Massachusetts
Lange, Frederick	UC Santa Barbara
Langer, Robert	MIT

EXPERT (VWC Organizer)	AFFILIATION
Larsen, John	Penn State University
Laurencin, Cato	University of Virginia
Leal, Gary	UC Santa Barbara
Lee, Kelvin	Cornell University
Lee, L. James	Ohio State University
Lee, Sang Yup	KAIST
Lee, Vincent	FDA
Lesko, Jack	Virginia Polytechnic Institute
Lewis, Jennifer A.	University of Illinois at Urbana Champaign
Liao, James	UCLA
Linninger, Andreas	University of Illinois
Lips, Alexander	Unilever
Liu, Jun	PNNL
Loy, Doug	University of Arizona
Luss, Dan	University of Houston
Macosko, Chris	University of Minnesota
Madix, Robert	Stanford University
Maggioli, Victor	Feltronics Corp.
Mallapragada, Surya	Iowa State University
Mallouk, Tom	Penn State University
Malone, Michael	University of Massachusetts
Maranas, Costas	Penn State University
Marinan, Mark	Dow Chemical
Mark, J.E.	University of Cincinnati
Marlin, Tom	McMaster University
Marquardt, Wolfgang	RWTH-Aachen
Marrucci, Guiseppe	University of Naples (Italy)
McAvoy, Tom	University of Maryland
McCarty, Perry L.	Stanford University
McCormick, Alon V.	University of Minnesota
McLeish, TCB	University of Leeds (UK)
Meyer, Anne	SUNY Buffalo
Michaels, James N.	Merck and Co.
Mikos, Antonios	Rice University
Mitragotri, Samir	UC Santa Barbara
Mooney, David	Harvard University
Morari, Manfred	ETH Zurich (Switzerland)
Mortensen, Andreas	Swiss Federal Institute of Technology
Mudan, Krishna	MSA Risk Consulting
Narasimhan, Balaji	Iowa State University
Nauman, Bruce	RPI
Ni, Xiong-Wei	Heriot-Watt University (UK)
Nielsen, Jens	Technical University of Denmark
Nienow, Alvin	University of Birmingham
Norris, David	University of Minnesota
Ober, Chris	Cornell University
Ogunnaike, Tunde	University of Delaware

EXPERT (VWC Organizer)	AFFILIATION
Okano, Teruo	Tokyo Women's Medical College (Japan)
Overton, Tim	Dow Chemical
Ozin, Geoffrey	University of Toronto (Canada)
Palsson, Bernhard	UC San Diego
Panagiotopoulos, Athanassios	Princeton University
Pandis, Spyros	Carnegie Mellon University
Papoutsakis, Terry	Northwestern University
Paul, Don	University of Texas-Austin
Pearson, Ray	Lehigh University
Pekny, Joe	Purdue University
Pendergast, John Jr.	Dow Chemical
Penlidis, Alexander	University of Waterloo
Peper, Jody	University of Minnesota
Peppas, Nicholas	University of Texas-Austin
Pereira, Carmo	DuPont
Petrie, Jim	University of Sydney (Australia)
Pistikopoulos, Stratos	Imperial College (UK)
Ponton, Jack	University of Edinburgh (Scotland)
Pratsinis, Sotiris E.	ETH Zurich (Switzerland)
Prausnitz, John	LBNL/UC Berkeley
Prud'homme, Robert	Princeton University
Rao, Govind	University of Maryland-Baltimore
Ray, W. Harmon	University of Wisconsin
Register, Richard	Princeton University
Reklaitis, Gintaras	Purdue University
Richon, Dominque	CEP/TEP, ENSMP (France)
Rochelle, Gary	University of Texas-Austin
Russel, William	Princeton University
Russell, Alan	University of Pittsburgh
Sandler, Stan	University of Delaware
Schaak, Raymond	Texas A&M University
Schaffer, David	UC Berkeley
Schowalter, William	Princeton University
Schuth, Ferdi	MPI für Kohlenforschung (Germany)
Scranton, Alec	University of Iowa
Seal, Sudipta	University of Central Florida
Seborg, Dale	UC Santa Barbara
Sefton, Michael	University of Toronto (Canada)
Sehanobish, Kalyan	Dow Automotive
Seinfeld, John	California Institute of Technology
Shafi, Asjad	Dow Chemical
Shah, Nilay	Imperial College (UK)
Shirtum, Page	RPS Engineering
Shuler, Mike	Cornell University
Siddall, Jon	Dow Chemical
Sidkar, Subhas	NMRL, EPA
Sierka, Raymond	University of Arizona

EXPERT (VWC Organizer)	AFFILIATION
Siirola, Jeff	Eastman Chemical Co.
Sinclair Curtis, Jennifer	University of Florida
Smith, Philip	University of Utah
Smith, Robin	University of Manchester
Spannangel, Mary Anne	University of Illinois
Stephanopoulos, Gregory	MIT
Stone, Howard	Harvard University
Stadther, Mark	University of Notre Dame
Stucky, Galen	UC Santa Barbara
Stupp, Sam	Northwestern University
Sundaresan, Sankaran	Princeton University
Teja, Amyn	Georgia Tech
Tester, Jefferson	MIT
Thibodeaux, Louis	Louisiana State University
Tirrell, David	California Institute of Technology
Tirrell, Matthew	UC Santa Barbara
Towler, Gavin	UOP (USA)
Vaia, Rich	AFRL
Varma, Arvind	Purdue University
Vayenas, Constantinos G.	University of Patras
Velev, Orlin	North Carolina State University
Virkar, Anil	University of Utah
Wall, Terry	University of Newcastle
Wandrey, Christian	Institute of Biotechnology (Germany)
Wang, Danny	MIT
Wang, Zhen-Gang	California Institute of Technology
Wassick, John	Dow Chemical
Webb, Colin	University of Manchester
Weber, W.J. Jr.	University of Michigan
Wei, James	Princeton University
Weinberg, W. Henry	UC Santa Barbara
Weitz, David	Harvard University
Wender, Irving	University of Cape Town (South Africa)
Wendt, Jost	University of Utah
West, David	Dow Chemical
West, Jennifer	Rice University
Westerberg, Arthur	Carnegie Mellon University
Westmoreland, Phillip	University of Massachusetts
White, Ralph E.	University of Southern California
Whitesides, George	Harvard University
Wilson, Grant	University of Texas-Austin
Winey, Karen	University of Pennsylvania
Wittrup, Dane	MIT
Xia, Younan	Washington University
Yager, Paul	University of Washington
Yang, Hong	University of Rochester
Yang, Ralph	University of Michigan

EXPERT (VWC Organizer)	AFFILIATION
Zaks, Alex	Schering-Plough
Zasadzinski, Joseph	UC Santa Barbara
Zhao, Huimin	University of Illinois
Zheng, Zhipling	University of Arizona

APPENDIX 3B

The List of Journals Examined for Publications and Citations

No.	Journal	2005 Impact Factor
Journals with Broad Coverage of Sciences and Engineering		
1	Science	30.927
2	Nature	29.273
3	Proceedings of the National Academy of Science	10.231
4	Physical Review Letters	7.489
5	Journal of the American Chemical Society	7.419
Journals with Broad Coverage of Chemical Engineering Research		
6	AIChE Journal	2.036
7	Chemical Engineering Science	1.735
8	Industrial and Engineering Chemistry Research	1.504
9	Chemical Engineering Research and Design	0.792
10	Canadian Journal of Chemical Engineering	0.574
11	Chemie Ingenieur Technik	0.392
Area-1: Engineering Science of Physical Processes		
12	Journal of Physical Chemistry B	4.033
13	Journal of Chemical Physics	3.138
14	Journal of Membrane Science	2.654
15	Journal of Rheology	2.423
16	Journal of Fluid Mechanics	2.061
17	Journal of Colloid and Interface Science	2.023
18	Separation and Purification Technology	1.752
19	Separation and Purification Review	1.571
20	Granular Matter	1.517
21	Fluid Phase Equilibria	1.478
22	Rheologica Acta	1.432
23	Journal of Chemical Thermodynamics	1.398
24	Molecular Simulation	1.345
25	International Journal of Multiphase Flow	1.306
26	Journal of Non-Newtonian Fluid Mechanics	1.268
27	Powder Technology	1.219
28	Separation Science and Technology	0.834
Area-2: Engineering Science of Chemical Processes		
29	Angewandte Chemie (International Edition)	9.596
30	Journal of Catalysis	4.780
31	Macromolecules	4.024
32	Applied Catalysis-B	3.809
33	Journal of Polymer Science Part A: Polymer Chemistry	3.027
34	Journal of Power Sources	2.770
35	Applied Catalysis-A	2.728

No.	Journal	2005 Impact Factor
36	Electrochimica Acta	2.453
37	Catalysis Today	2.365
38	Journal of the Electrochemical Society	2.190
39	Solid State Ionics	1.571
40	Journal of Applied Electrochemistry	1.282
41	International Journal of Chemical Kinetics	1.188
42	Surface and Interface Analysis	0.918
43	Journal of Polymer Engineering	0.312
44	Studies in Surface Science and Catalysis	0.307
Area-3: Engineering Science of Biological Processes		
45	Nature Biotechnology	22.738
46	Bioinformatics	6.019
47	Proteins: Structure, Function, and Bioinformatics	4.684
48	Applied and Environmental Microbiology	3.818
49	Protein Science	3.618
50	Metabolic Engineering	2.484
51	Biotechnology & Bioengineering	2.483
52	Biotechnology Progress	1.985
53	Process Biochemistry	1.796
54	Enzyme and Microbial Technology	1.705
55	Bioprocess and Biosystems Engineering	0.807
Area-4: Molecular and Interfacial Science and Engineering		
56	Journal of Physical Chemistry B	4.033
57	Langmuir	3.705
58	Journal of Colloid and Interface Science	2.023
59	Colloids and Surfaces B	1.588
60	Colloids and Surfaces A	1.499
Area-5: Materials		
61	Progress in Polymer Science	16.045
62	Nature Materials	15.941
63	Nano Letters	9.847
64	Advanced Materials	9.107
65	Advanced Functional Materials	6.770
66	Chemistry of Materials	4.818
67	Inorganic Chemistry	3.851
68	Acta Materialia	3.430
69	Polymer	2.849
70	Composites Science and Technology	2.184
71	Journal of Materials Research	2.104
72	Journal of Polymer Science Part B: Polymer Physics	1.739
73	Journal of the American Ceramic Society	1.586
74	Journal of the European Ceramic Society	1.567
75	Materials Research Bulletin	1.380

No.	Journal	2005 Impact Factor
76	Polymer Engineering and Science	1.076
77	Composite Structures	0.953
78	Journal of Materials Science	0.901
79	Journal of Ceramic Society of Japan	0.749
80	Polymer Composites	0.628
81	Inorganic Materials.	0.387
82	Journal of Polymer Engineering	0.312

Area-6: Biomedical Products and Biomaterials

83	Advanced Drug Delivery Reviews	7.189
84	Biomaterials	4.698
85	Journal of Controlled Release	3.696
86	Biomacromolecules	3.618
87	Journal of Orthopaedic Research	2.916
88	Tissue Engineering	2.887
89	Pharmaceutical Research	2.752
90	Journal of Biomedical Materials Research	2.743
91	European Journal of Pharmaceutical Sciences	2.347
92	Annals of Biomedical Engineering	1.997
93	Journal of Biomaterials Science, Polymer Edition	1.409
94	Journal of Materials Science: Materials in Medicine	1.248

Area-7: Energy

95	Carbon	3.419
96	Progress in Energy and Combustion Science	3.371
97	Combustion and Flame	2.258
98	Solar Energy Materials and Solar Cells	2.002
99	Fuel	1.674
100	Energy and Fuel	1.494
101	Fuel Process Technology	1.171
102	SPE Journal	0.816
103	Combustion Science and Technology	0.774
104	Proceedings Combustion Institute	0

Area-8: Environmental Impact and Management

105	Environmental Science and Technology	4.054
106	Atmospheric Chemistry and Physics	3.495
107	Water Research	3.019
108	Journal of Geophysical Research	2.784
109	Atmospheric Environment	2.724
110	Tellus B	2.592
111	Journal of Aerosol Science	2.477
112	Environmental Toxicology and Chemistry	2.414
113	Chemosphere	2.297
114	Journal of Atmospheric Science	2.078
115	Water Resources Research	1.939

No.	Journal	2005 Impact Factor
116	Aerosol Science and Technology	1.935
117	Journal of Contaminant Hydrology	1.733
118	Ground Water	1.419
119	Journal of Nanoparticle Research	1.699
120	Journal of the Air and Waste Management Association	1.317
121	Ecological Economics	1.179
Area-9: Process Systems Development and Engineering		
122	INFORMS Journal on Computing	1.762
123	Automatica	1.693
124	SIAM Journal on Scientific Computing	1.509
125	Computers & Chemical Engineering	1.501
126	Mathematical Programming	1.497
127	Journal of Process Control	1.433
128	SIAM Journal on Optimization	1.238
129	Chemical Engineering and Processing	1.159
130	Computational Optimization and Applications	0.886
131	Chemical Engineering Research & Design	0.792
132	Chemical Engineering and Technology	0.678
133	Journal of Global Optimization	0.662
134	Journal of Optimization Theory and Applications	0.612
135	Annals of Operations Research	0.525
136	Process Safety Progress	0.320
137	Optimization and Engineering	

4

Benchmarking Results:
Detailed Assessment of U.S. Leadership
by Area of Chemical Engineering

Chapter 3 provided an assessment of U.S. chemical engineering research at large. In this chapter we focus the assessment on each area/subarea of chemical engineering research. Based on the analysis of data regarding the composition of the VWC, publications and citations, patents, recognition of individual researchers through prizes and awards, and prevailing trends, the Panel compiled an overall assessment for each subarea in terms of the following two indices:

- Current Position of U.S. Research in Chemical Engineering
- Expected Future Position of U.S. Research in Chemical Engineering

In assessing the future position of U.S. chemical engineering research the Panel took into consideration, in addition to the above, a set of key determinants of leadership, such as the following:

- intellectual quality of researchers and ability to attract talented researchers
- maintenance of strong, research-based graduate educational programs
- maintenance of strong technological infrastructure
- cooperation among government, industrial, and academic sectors
- adequate funding of research activities

Table 4.45 summarizes the Panel's assessment of the Current and Future Positions of U.S. Research Chemical Engineering in all subareas alongside the expected future trends.

The leadership determinants (ability to attract talented students, educational and research programs, technological infrastructure, cooperation among government, industry, and academia, and funding) and their (projection) are analyzed in Chapter 5.

4.1 AREA-1: ENGINEERING SCIENCE OF PHYSICAL PROCESSES

This area encompasses research in the science and engineering of processes, which are characterized, primarily, by physical phenomena. It has been divided into the following five subareas:

- transport processes
- thermodynamics
- rheology
- separations
- solid particle processes

4.1.a Transport Processes

The role of transport processes in chemical engineering has evolved from fundamental understanding and cutting edge/frontier research in the 1960s into two parallel fronts: one deepening fundamentals, the other evolving towards applications. It has also taken a role as a platform technology, with a presence in nearly all areas of chemical engineering, spanning from traditional processing (e.g., reactors, separation systems) to biological applications and materials. Transport phenomena, with or without chemical reaction, are at the heart of all processing systems at any scale (macro, micro, nano) and as such are at the very core of chemical engineering; indeed, in what may be a commonly held belief, they define chemical engineering.

In defining the scope of this subarea we have considered traditional aspects of fluid mechanics, such as low Reynolds number flows and turbulent flows including multiphase flows; fluid-particle systems; all types of mass and heat transport, including chemically assisted mass transport; flows of complex fluids (connecting smoothly with rheology); flows induced by electric or magnetic fields (bridging with colloidal science); and transport at interfaces. Other aspects include a blend of research and practical considerations, such as numerical simulation for analysis and design as well as prediction of and correlations for transport properties. Topics of current importance have evolved towards fluid mechanics and mass transport at interfaces and small scales, as in microfluidics, nanoscale devices, molecular-level modeling of tribology, and biological molecules and living cells. Particulate and multiphase flows, interfacial flows, non-Newtonian fluid mechanics, and flow mechanics of complex fluids and biomolecules remain subjects of intense research interest due to their intellectual challenge and broad range of potential applications.

U.S. Position. The number of experts for the VWC in this area was five U.S. and two non-U.S. The percentage of U.S. participants in the Virtual Congress was 81% when multiple entries for the same person were allowed. This was among the highest representations in all subareas considered by this panel. The percentage was 77% when name duplication was disallowed, and indicates that several U.S. names appeared in multiple lists. These numbers point to strong U.S. leadership in transport. An analysis of names reveals that a significant number of the names are associated with "classical" fluid mechanics as opposed to mass transfer or energy transfer, which are clearly regarded today as mature areas in chemical engineering.

A survey of the flagship journals in the fluid mechanics area, the *Journal of Fluid Mechanics* and *Physics of Fluids* reveals that the number of U.S. contributions, across disciplines, from 1990 to 2006, increased by a factor of 2, but its relative percentage was reduced by 9%, due to higher rates from, European Union (EU) and Asia (see Table 4.1). In terms of quality and the impact, U.S. contributions dominate (66%) the list of the 50 most-cited papers (Table 4.2).

The chemical engineering contributions worldwide have more than doubled in number, maintaining roughly the same relative percentage, about 8% of the total papers.

U.S. chemical engineering has dominated the chemical engineering contributions: 84% in the period 1990-1994 and 75% in the period of

TABLE 4.1 Publications in *Journal of Fluid Mechanics* and *Physics of Fluids*

	1990-1994		1995-1999		2000-2006	
		%		%		%
Total Number of Papers	2,070		3,439		5,029	
Total No. of U.S. Papers	1,174	57	1,836	53	2,300	46
Total No. of Chem. Eng. Papers	163	7.87	286	8.32	389	7.74
U.S., Chem. Eng.	143	87.73	245	85.66	289	74.29
EU, Chem. Eng.	6	3.68	19	6.64	48	12.34
Asia, Chem. Eng.	12	7.36	33	11.54	53	13.62
Canada, Chem. Eng.	7	4.29	9	3.15	9	2.31
S. America, Chem. Eng.	1	0.61	0	0.00	4	1.03
Internationalization (overlap)		3.68		6.99		3.60

TABLE 4.2 Distribution of the 50 Most-Cited Papers in *Journal of Fluid Mechanics* and *Physics of Fluids*

	1990-1994	1995-1999	2000-2006
No. of U.S. Papers	36	30	33
No. of Chem. Eng. Papers	3	3	3
No. of U.S. Chem. Eng. Papers	3	2	3
(% share among chemical engineering papers)	(100%)	(66%)	(100%)

2000-2006. In addition, all papers from chemical engineering researchers in the list of 50 most-cited papers for 2000-2006 come from the United States. These numbers are consistent with the dominant representation of U.S. chemical engineers in the Virtual World Congress for this subarea, and both indicate that the relative U.S. position is "Dominant, at the Forefront," in relationship to chemical engineering research elsewhere in the world. The real competition comes from other disciplines, notably physics, applied mathematics, and mechanical engineering. Indeed, Table 4.2 indicates that only 6% of the 50 most-cited papers come from U.S. chemical engineering research. This is primarily due to the fact that chemical engineering research activities in fluid mechanics represent a small subset of this field.

Analysis of the publications from mainstream journals of chemical engineering such as *AIChE Journal*, *I&EC Research* and *Chemical Engineering Science* indicates that in 1995 there were about 1.5 papers from U.S. authors for every paper from a non-U.S. author. This ratio has changed to about 0.5 to 0.6, following the significant increase in the research output from the European Union and Asia. It should be noted that a number of publications that in the past would have gone to classical journals, such as the *Journal of Fluid Mechanics* and *Physics of Fluids*, now go to more peripheral publications associated with niche areas, e.g., microfluidics. At the same time there has been a decrease in the number of publications in once classical and central areas of chemical engineering, such as two-phase flow, heat transfer, fluidization, and the like. The volume of research in these areas and the number of ensuing publications from Asian countries has increased substantially.

Relative Strengths and Weaknesses. U.S. chemical engineering scholarly activities in transport phenomena have, until the mid 1980s, attracted some of the best talent in the United States and transport was considered to be a prestige area. Now, opportunities for long-range funding in pure fluid mechanics and fundamentals in mass transport are virtually nonexistent in the United States. This can have long-term negative consequences for chemical engineering in the United States. Loss of transport strength will result in a loss of differentiation that has been critical for chemical engineering work across multiple areas, at a time when processing at the micro- and nano scale, formation of structured materials and their processing into a multitude of functional parts, the production of efficient energy devices (at any scale), and efficacy of a broad range of biomedical devices, may hinge on better understanding of the associated transport processes.

With the exception of niche centers such as Stanford's Center for Turbulence Research (largely dominated by mechanical engineering), the United States has surprisingly few large institutes wholly dedicated to fluid mechanics and mass transport, and even fewer with a significant component of chemical engineering. (There are, however, a few devoted to mixing, for example). Current U.S. research in transport tends to be concentrated on applications in fluid mechanics. Fluid mechanics in industry is dispersed throughout many areas, though recognizable pockets may include computational fluid mechanics, e.g., analysis and design of reactors with complex flows, heat and mass transport, as well as groups focused on fluid mechanics of suspensions, high-precision coating processes, mixing, and transport and reaction in heterogeneous and porous systems.

Future Prospects. As the framework of transport phenomena developed and tools were created there was a migration outwards, and many areas that were once frontiers of transport research have become permanently integrated with many surrounding areas. It has become increasingly difficult to delineate the boundaries between transport and colloidal science, transport and solid/particulate systems, and transport and rheology.

Significant advances have taken place over the last decade. Some of these advances have been in traditional areas such as simulation of multiphase and turbulent flows at single and multiple length and time scales, mixing, and coating flows. It is now possible to simulate efficiently suspensions and emulsions, and flow of non-Newtonian fluids for almost any admissible constitutive law, and to apply fundamental transport phenomena to a variety of practical microfluidic devices.

New areas and opportunities lie at the intersection of transport and colloid science, e.g., cases involving sophisticated couplings of interparticle colloidal forces, and external fields and fluid mechanics. These ideas find application in the directed self-assembly of materials and separations as in electrophoresis, diffusiophoresis, induced-charge electrophoresis, and others. New challenges are arising in microrheology, as it is used to probe complex fluids and biological systems. The frontier areas of designing and making nanocomposites and nanoparticulate/polymer complex fluids require the simultaneous tailoring of transport, rheological, and mechanical properties. Another area of active research involves granular matter and study of jamming, ageing and flow properties of glassy/disordered materials ranging from pastes to polymers to granular media.

Panel's Summary Assessment. The current U.S. position is at the "Forefront," and the Panel expects that in the future it will be "Among World Leaders."

4.1.b Thermodynamics

Thermodynamics has evolved from the classical studies of estimating thermophysical properties and phase behavior of fluids that defined the field in the middle of the 20th century to a much more molecular and science-based field with a significantly broader range of applications. Experimental studies now examine new formulations of consumer products, e.g., refrigerants; new solvents as diverse as carbon dioxide and ionic liquids; degradation and stabilization of biological molecules, e.g., proteins, DNA, RNA; supercooled liquids and glasses; thermophysical properties of biological systems; structure and properties of polymers and blends; nucleation and growth; and others. Theoretical advances frequently follow application of the principles of statistical thermodynamics and, increasingly, quantum mechanics, to engineering problems. Molecular simulations are becoming quite entrenched and their predictive efficiency is progressing by leaps and bounds. Examples include improvements in the understanding of the properties of water; *ab initio* calculations of molecular interactions important in biological processes, e.g., complex immune systems, and estimation of thermophysical properties and phase behavior of biomolecules; computational studies of self-assembled systems at meso- and nano scale, e.g., copolymers, polymer blends, composites; theoretical and computational studies on nucleation/formation and growth of e.g., colloids, crystals, emulsions, foams;

Thermodynamics is an integral part of the chemical engineering science base and underlies many traditional chemical engineering unit operations. As the academic interests and industrial emphasis have been shifting towards better understanding of molecular-level phenomena, thermodynamics is playing a key role in advance understanding of the molecular forces underlying molecular organization, self-assembly, and materials design, and in developing new media and their applications, such as environmentally benign solvents for dry cleaning; water-based dispersions of inks, dyes, and pigments for the electronic and automobile industries; and functional structured fluids for the personal care industry, home and office products industry, food industry, and other sectors.

U.S. Position. In addition to the mainstream chemical engineering journals, such as *AIChE Journal, I&EC Research and Chemical Engineering Science* other principal journals in this subarea include the *Journal of Chemical Physics*, *Journal of Physical Chemistry B*, *Molecular Simulations*, *Fluid Phase Equilibria*, and the *Journal of Chemical Thermodynamics*. In the first three journals, the relative contribution of U.S. chemical engineering researchers against non-US contributions has decreased from 3.5 U.S. papers per non-U.S. paper to about 1.0. Significant increases in submissions from European Union and Asian countries have been the main factor contributing to this change. In each of the latter five journals the U.S. contributions in the past few years across disciplines range from 15% to 40%, and in all cases contributions from the European Union are more numerous than those from the United States (Tables 4.3 and 4.4).

Tables 4.5, 4.6, and 4.7 summarize the trends in chemical engineering contributions for the five journals. The numbers indicate that for the past 20 years chemical engineering papers have captured a roughly constant relative percentage of all publications, ranging from 4% to 30%. For the *Journal of Chemical Physics* and *Journal of Physical Chemistry B* the percentage contribution from chemical engineers worldwide has been increasing (from about 4% in 1990-1994 to over 7% in 2000-2006), but it remains at low levels, with contributions from chemists outnumbering those of chemical engineers by factors of 3 to 6. Tables 4.6 and 4.7 also indicate that the percentage contribution of U.S. chemical engineers has been decreasing over the past 20 years, e.g., from 40% to 23% (combined numbers

TABLE 4.3 Publications in Three Area-Specific Journals for Thermodynamics

	J. of Chemical Physics			J. of Physical Chemistry-B			Molecular Simulations		
	% U.S.	U.S. Papers	EU Papers	% U.S.	U.S. Papers	EU Papers	% U.S.	U.S. Papers	EU Papers
2003	40	1064	1311				30	27	35
2004	39	1048	1363	36	857	1093	29	23	46
2005	40	1110	1400	36	1169	1317	25	33	53

TABLE 4.4 Publications in Two Area-Specific Journals for Thermodynamics

	Fluid Phase Equilibria			J. of Chemical Thermodynamics		
	% U.S.	U.S. Papers	EU Papers	% U.S.	U.S. Papers	EU Papers
2003	12	27	95	18	31	45
2004	19	50	107	16	21	25
2005	15	58	134	15	22	37

TABLE 4.5 Publication Trends in *Journal of Chemical Physics* and *Journal of Physical Chemistry-B*

	1990-1994		1995-1999		2000-2006	
		%		%		%
Total Number of Papers	9,672		15,582		30,064	
No. of U.S. Papers	5,516	57.00	6,936	45.00	11,819	39.00
No. of Chem. Eng. Papers	369	3.82	866	5.56	2,182	7.26

TABLE 4.6 Publication Trends in *Fluid Phase Equilibria* and *Journal of Chemical Thermodynamics*

	1990-1994		1995-1999		2000-2006	
		%		%		%
Total Number of Papers	1,714		2,102		2,630	
No. of U.S. Papers	430	25.00	432	21.00	439	17.00
No. of Chem. Eng. Papers	478	27.89	621	29.54	757	28.78
U.S., Chem. Eng.	191	39.96	209	33.66	178	23.51
EU, Chem. Eng.	83	17.36	115	18.52	166	21.93
Asia, Chem. Eng.	162	33.89	223	35.91	279	36.86
Canada, Chem. Eng.	52	10.88	49	7.89	44	5.81
S. America, Chem. Eng.	3	0.63	23	3.70	39	5.15

TABLE 4.7 Publication Trends in *Molecular Simulations*

	1990-1994		1995-1999		2000-2006	
		%		%		%
Total Number of Papers	159		220		496	
No. of U.S. Papers	48	30.00	36	16.00	125	25.00
No. of Chem. Eng. Papers	26	16.35	19	8.64	74	14.92
U.S., Chem. Eng.	25	96.15	14	50.00	42	56.76
EU, Chem. Eng.	1	3.85	7	25.00	14	18.92
Asia, Chem. Eng.	1	3.85	6	21.40	23	31.08
Canada, Chem. Eng.	1	3.85	0	0.00	2	2.70
S. America, Chem. Eng.	0	0.00	1	3.60	1	1.35

for *Fluid Phase Equilibria* and *Journal of Chemical Thermodynamics*), and from 96% to 57% (*Molecular Simulations*). These reductions are primarily due to higher growth rates in other parts of the world, notably European Union and Asia.

In terms of quality and impact, Table 4.8 summarizes the distribution of the 50 most-cited papers for three groups of journals. Overall, across

TABLE 4.8 Distribution of the 50 Most-Cited Papers for Combined *Journal of Chemical Physics* and *Journal of Physical Chemistry-B*, Combined *Fluid Phase Equilibria* and *Journal of Chemical Thermodynamics* and *Molecular Simulations*

	J. of Chemical Physics and J. of Physical Chemistry B			Fluid Phase Equilibria and J. of Chemical Thermodynamics			Molecular Simulations		
	1990-1994	1995-1999	2000-2006	1990-1994	1995-1999	2000-2006	1990-1994	1995-1999	2000-2006
No. of U.S. Papers	32	28	31	20	21	11	12	4	11
No. of Chem. Eng. Papers	2	1	4	35	31	32	7	6	9
No. of U.S. Chem. Eng. Papers	2	1	4	17	13	7	5	3	8
(% share among chemical engineering papers)	(100%)	(100%)	(100%)	(50%)	(42%)	(28%)	(70%)	(50%)	(89%)

disciplines, the United States possesses a position "Among the Leaders" with strong competition from the European Union. With respect to chemical engineering contributions, the United States is in a "Dominant, at the Forefront" position. Furthermore, when we take a close look at the list of the 100 most-cited papers (2000-2006) in chemical engineering at large, we notice that the field of thermodynamics is well represented in the list—two of the top three and three of the top five most-cited papers have thermodynamics as their subject. Pioneering papers in this field are published in top journals, including *Nature, Science,* and *Proceedings of the National Academy of Sciences.*

Two hundred seventeen participants were identified in the area of thermodynamics for the Virtual World Congress, and 68% of them were from the United States (61% when duplications were disallowed), which is about the same for the overall U.S. participation in the Virtual World Congress. This is in line with the numbers and impact of U.S. publications among chemical engineering researchers, firming up the conclusion that U.S. chemical engineering research in the area of thermodynamics is "Dominant, at the Forefront."

Relative Strengths and Weaknesses. Thermodynamics is a large and vibrant field in chemical engineering worldwide, and like many engineering fields that are closely linked to science, many significant contributions come from workers outside of chemical engineering departments. This is true in the United States, but is particularly evident in the European Union. Thus, it is difficult to benchmark only U.S. chemical engineers in this arena, and there are many substantial and important U.S. academic collaborations that involve chemical engineers together with chemists or physicists, all working on both experimental and theoretical problems. This interdisciplinary work is clearly a strength, and it allows new ideas to be readily applied to problems of interest to chemical engineering practitioners.

The field has in general expanded over the time period represented in our analysis of publications, with growth by a factor of 2 in publications in *Fluid Phase Equilibria* and a factor of over 4 in *Molecular Simulations.* These journals primarily reflect, respectively, reports of experimental and simulation studies. The relative rates of growth indicate a substantially larger rate of new developments in simulations, which is of course to be expected given the availability of increasing computational power. The percentage of U.S. contributions in *Fluid Phase Equilibria* was 20% in 1997 and 15% in 2005, and in *Molecular Simulations* was 10% in 1997 and 25% in 2005. This further validates the impression that simulations are a more attractive area for research than are traditional experiments, but in all cases the absolute number of U.S. publications increased.

Participants in the Virtual World Congress spanned a range of ages with a healthy distribution of experience. There are certainly senior leaders, but also a good diversity of younger chemical engineers involved in this area, particularly on the computational and simulation side. The relative lack of experimental activity may be worrisome for the future prospects.

From the publication analysis there appears to be no dramatic shift in the international distribution of articles with the United States and EU a substantial majority, but there is a noticeable increase of papers from China in *Fluid Phase Equilibria*.

Future Prospects. Significant advances in molecular simulation for the estimation of thermodynamic properties have taken place during the past 10 years for complex systems such as polymers, surfactants, liquid crystals, subcooled water, biomacromolecules, and ionic liquids. Application of quantum mechanics for the calculation of intermolecular forces in phase equilibrium description of fluid mixtures, elucidation of the effects of pressure and solutes on the thermodynamics of hydrophobic hydration of large and small solutes, development of ionic liquids as solvents for separations, and thermodynamics of glasses and disordered systems are some of the other major advances in recent years.

Thermodynamics will continue to be a critical area of chemical engineering for the foreseeable future, and a large and continually growing portion of the field will continue to exploit computer simulations to address practical problems. Other areas of growth will be the application of thermodynamics to biological and complex materials synthesis and processing problems. These will occur both in processing steps in industry, where for example thermodynamic studies can guide optimization of unit operations such as protein crystallization, and in increasingly sophisticated descriptions of the molecular interactions responsible for recognition events. As the theoretical tools in this field enable more accurate descriptions of molecular features, experiments will also probe finer scales. This is particularly true in applications of thermodynamics to descriptions of self-assembly processes involving surfactants, polymers and polyelectrolytes, and other nanoscale building blocks, which are becoming the core components for high added-value products in a variety of industries.

The Panel expects that in the future the following items will continue to attract the research interest: multiscale modeling, which starts with atomic-level descriptions for the simulation of soft matter-colloid solutions, surfactants and micellar solutions, polymers, and self-assembly from these; thermodynamics of biological molecules, the hydrophobic effect, and protein folding; thermodynamics of solubility, bioavailability, and protein binding for drug discovery and development; thermodynamics of small systems needed for the simulation and design of nanostructures; nontraditional measurements (e.g., spectroscopy, X-ray diffraction) to obtain equations of state and other models; practical thermodynamic models needed for product engineering and green products and processes.

Panel's Summary Assessment. The current U.S. position is "Among World Leaders," and in the near future it will remain so.

4.1.c Rheology

In its broadest definition, rheology is the study of deformation, and ultimately flow, of any material under the influence of an applied force or stress. The study of rheology and the application of its teachings are important to the development of both chemical processes and products. Rheology occupies the area between solid mechanics, which is not usually the realm of chemical engineers, and Newtonian fluid mechanics, which is. As a result many chemical engineers are involved in at least some aspect of rheology research, and most of the leading chemical engineering departments have at least one rheologist. Current research directions include emphasis on fluids in microfluidic flows, flows of complex fluids, and flows in biological systems. The modern study of microrheology, for example, is shedding light on the formation of actin gels, and rheology can be used to probe the kinetics of nanoparticle formation. Rheology will be of increasing importance in the design and characterization of both food and personal care products—the formulation of both is now increasingly based on science rather than empiricism.

U.S. Position. For the Virtual World Congress the number of experts in this area was eight, with six from the US. 113 speakers were identified for the Virtual World Congress, with 62% of them from the United States when multiple nominations were included. This share was only slightly below the 67% of the overall U.S. participation in the Virtual World Congress participation. The percentage of the U.S. participation in the Virtual World Congress drops to 52% when no duplications were allowed, indicating that the U.S. research in this subarea is "Among the Leaders," but not dominant.

In addition to the three mainstream chemical engineering journals, principal journals in the field include the *Journal of Rheology*, *Journal of Non-Newtonian Fluid Mechanics*, and *Rheological Acta*. U.S. and European Union contributions are about equal in number except for the *Journal of Non-Newtonian Fluid Mechanics*, which is dominated by European Union papers (Table 4.9). There is a negligible contribution of papers from China and India to the journals surveyed. The percentage of U.S. authorship has been relatively constant over the past 10 years. For all three journals, authors affiliated with chemical engineering departments are the dominant

TABLE 4.9 Geographic Distribution of Publications in *Journal of Rheology, Journal of Non-Newtonian Fluid Mechanics,* and *Rheological Acta*

	Journal of Rheology			Journal of Non-Newtonian Fluid Mechanics			Rheology Acta		
	% U.S.	Papers U.S.	Papers EU	% U.S.	Papers U.S.	Papers EU	% U.S.	Papers U.S.	Papers EU
1997	51	38	28	32	33	61	52	33	22
2000	32	25	43	44	50	43	33	25	40
2003	48	41	44	32	31	48	25	15	29
2004	39	30	40	33	38	60	19	13	41
2005	53	42	38	21	27	80	33	26	36

TABLE 4.10 The Percentage Contributions of Researchers from Chemical Engineering in *Journal of Rheology, Journal of Non-Newtonian Fluid Mechanics,* and *Rheological Acta*

	Journal of Rheology	Journal of Non-Newtonian Fluid Mechanics	Rheology Acta
1997	27	21	32
2000	35	37	24
2003	38	25	31
2004	43	30	24
2005	53	19	30

contributors (Table 4.10), so on balance U.S. chemical engineers contribute about one-third of the papers published in this field. At least three rheological papers appear on the list of the 100 most-cited papers in chemical engineering (2000-2006), although in *Macromolecules*, not in the journals listed above. The percentage of U.S. paper contributions is substantially lower than the U.S. participation in the Virtual World Congress. However, when we consider only chemical engineering researchers, publications and Virtual World Congress participation come closer in agreement.

Relative Strengths and Weaknesses. Rheology is a relatively small field worldwide. The number of publications has been constant over time, so the field is relatively stagnant in that sense, but rheological ideas are now being applied to many new areas including complex fluids and biological assemblies. It is nearly certain that the results of those studies are in some cases being published outside of the traditional rheological journals, so this publication analysis does not measure the growth of these new areas of application.

Future Prospects. Rheology has been and will continue to be an area of significant chemical engineering research for the foreseeable future, but is unlikely to grow. Rheologists will continue to expand the scope of their work to new problems and applications. This is particularly true in applications of rheology to characterize and control self-assembly processes involving surfactants, polymers and polyelectrolytes, and other colloidal or nanoscale building blocks. The rate of this growth could be expanded, and the rheology community should embrace new interactions with industry as, for example, established by the International Fine Particle Research Institute (IFPRI).

Panel's Summary Assessment. The current U.S. position is at the "Forefront/Among World Leaders," and in the near future, it will be "Among World Leaders."

4.1.d Separations

Separation is critical to every chemical process and, typically, more than half of the invested capital in a plant is dedicated to separation and purification. There is a wide variety of unit operations employed, including:

- concentration-controlled separations: absorption, adsorption, distillation, drying, etc.
- electric and/or magnetic field-controlled separations: electrostatics, electrophoresis, electroosmosis, etc.
- gravity-controlled separations: centrifugation, liquid/liquid, liquid/gas, solid/gas, etc.
- size-controlled separations: membranes, sieves, etc.
- chemically assisted separations

Research activities cover most of these separation methods, and researchers have expanded the range of industrial problems to include separation and purification of bioproducts, novel membranes for fuel cells and water reuse, separations in microsystems, and others. Bioprocess-related separations are discussed in Section 4.3.c.

U.S. Position. The United States has a strong historical background in separations within the chemical process industries (CPI), but is not necessarily in a dominant position. The European community (particularly Germany) has had strong programs in applied technology, including separation technology. A survey of the key journals in this area (*Separation Processes, Separation Science and Technology, Filtration and Separation*) reveals that the U.S. contribution from 1997 to 2005 has remained relatively constant at 13%-30%.

This seems to underrepresent the U.S. status and may be a result of the journals selected for the study. Indeed, when we analyze the separations-oriented publications in the mainstream chemical engineering journals, *AIChE Journal*, *I&EC Research*, and *Chemical Engineering Science*, the ratio of U.S. to non-U.S. papers has changed from 60/40 (mid 1990s) to 50/50 (2000-2006), indicating a higher rate of growth for non-U.S. contributions, while the United States has remained the leader in numbers and impact.

To account for membrane separations, contributions in the *Journal of Membrane Science* were analyzed. The percentage of U.S. contributions declined from 29% (1990-1994) to 22% (1995-1999) to 20% (2000-2006). While the percentage of chemical engineering contributions worldwide remained about the same (35% to 38% of all published papers) from 1990-2006, the representation of U.S. chemical engineers declined from 49% of all chemical engineering contributions in 1990-1994 to 29% in 2000-2006. The percentage of the Asian chemical engineering contributions increased significantly during the same period from 29% in 1990-1994 to 46% in 2000-2006. Also, in terms of quality and impact we see an erosion of the U.S. position: 16 of the 30 most-cited papers in 1990-1994 were contributions from the United States and this number was reduced to 9 of the 30 most cited in 2000-2006. The difference was taken up by contributions from the European Union and Asia.

Furthermore, the Virtual World Congress suggested that 65% of the speakers would be from the United States and examination of U.S. patents granted in separations for 1995, 2000, and 2004 reveals that 55%, 53%, and 56%, respectively, were assigned to U.S. companies, suggesting a consistent and higher level of contribution.

Relative Strengths and Weaknesses. The United States has developed some successful industrial consortia such as Fractionation Research, Inc. (FRI), but these have usually a narrow focus (e.g., distillation) rather than a broad focus on separations. The Separations Research Program (SRP, University of Texas) is an exception. It has a broader focus that includes adsorption, extraction, and membranes. A unique area of U.S. strength is membrane research.

In general, interest in traditional chemical engineering unit operations, of which separations is a subset, is declining. There are few U.S. universities that offer graduate-level studies directed towards separations, compared to the past when separation research was present in virtually every U.S. chemical engineering department. The U.S.-based educators and industrial practitioners are an aging population. There is concern regarding training of future leaders in the area of process separations. For example, as the United States is seriously contemplating the biochemical production of ethanol as fuel from cellulose, it is the cost of ethanol separation and purification that could dominate the total process cost, and thus the economic viability of the proposal. Such prospects may be at risk, given the progressive deterioration in human and facilities infrastructure of the United States in this subarea.

There are few broad consortia in the United States that focus on larger topics such as process synthesis (of which separations is a critical subset) or hybrid separations development. There are active consortia in Europe and elsewhere that are looking into broad developments in chemical processing, which include separations. To give a concrete example, the concept of dividing wall columns (DWC) was strongly supported by a European consortium of companies and universities over the last several decades. The result is that European (particularly German) companies are more advanced in industrial implementation, even though the key patent work occurred in the United States.

Major advancements (as opposed to incremental) in separations technology are seriously hindered in the United States by the inability of academic and industrial partnerships to develop and test concepts on industrial chemicals of interest. Issues around intellectual property, scale and safety, and handling of hazardous chemicals in an academic environment, are serious obstacles to such partnerships.

Future Prospects. Realistically, one must describe our position in this area as challenging. Publications on separation synthesis, process intensification, and many other advanced topics are coming from Asia (particularly China) at an increasing pace. Asia and China in particular have the need and desire to develop a chemical process industry, and this need and desire is seen as an important part of that growth. The current escalation in the cost of energy and feedstocks should spark a renewed interest in improving separations methodology and technology. Many of the advancements require capital investment, and this has been seen as a roadblock in the past. As the price of energy rises, it is more likely that separations technologies that in the past could not support reinvestment could now support this expense. There are a number of advanced separations technologies that would require substantial research and validation, but there are also a number of technologies that could be implemented with only a modest research investment.

There should be a strong interplay between separations scientists and engineers and those working in thermodynamics and in the synthesis, characterization, and computational modeling of new separations membranes and other materials. Separations should grow to be considerably more multidisciplinary. The area of membrane separations will continue to be strong and thriving. The European Union is the primary competitor in terms of quality and impact (see numbers of cited papers, above). Asia is very competitive in membrane separations (see above) and will remain so in the near future. The number of Asian papers in membrane separations is larger than that from the United States and the quality and impact have increased significantly since 1990. Subjects of particular importance will be the development of efficient and selective processes for gas separation, the development of advanced techniques of solute separation, improved high-flux membranes, and low and high molecular weight solute separation. Ionic liquids for separations, merging Computational Fluid Dynamics with rate-based modeling of separation processes, and low-cost membrane systems for vapor-liquid contacting with significantly reduced energy requirements, are some of the technologies that will be developed in the near future. The United States will continue to play a primary role in this field mostly through the leadership of major U.S. chemical corporations. A focused dialog between academia and industrial practitioners is needed to counter new initiatives coming from Asia and Europe.

Panel's Summary Assessment. The current U.S. position is "Among World Leaders," and although in the future it is expected to weaken, it will still be "Among World Leaders."

4.1.e Solid Particles Processes

In this subarea we consider particle formation processes (nucleation, growth), particle measurement techniques (size, shape, distribution), processing (mixing, blending, and segregation), separation, attrition and agglomeration, compaction, sintering, tribology of particulate systems, and electrostatic effects in particle processing. The literature on the above topics is scattered among various branches of engineering—chemical, civil, and mechanical—as well as geophysics, pharmacy, and materials science. The most decidedly science-based work appears in physics. Current research has expanded the scope of engineering issues in this subarea to include formation, growth, scaling up, and processing of nanoparticles, but the discussion of the associated activities will be found in Section 4.5.d. Aerosols and the associated science and engineering aspects are discussed in Section 4.8.c.

U.S Position. For the Virtual World Congress the number of experts were five U.S., one non-U.S. The percentage of U.S. participants in the Virtual World Congress was 83% when duplication of names is allowed, and the percentage dropped to 51% when duplication was not allowed. This indicates that several U.S. names appeared in multiple lists. It is important to stress that the solids area was perceived by some to be in severe state of crisis in the United States as recently as a dozen years ago (see B. J. Ennis, J. Green, R. Davis, Legacy of Neglect in the United States, Chem. Eng. Progress, 90, 32-43, 1994.). There are indications that the picture is changing, and a few numbers from the Institute of Scientific Information are revealing. Consider the number of papers with the key words "particulate system," "granular material," "granular matter," and "granular flow," in the periods 1955-1996 and 1997-2006. The numbers are 168/69, 11/181, 530/1065, 143/624, respectively. The numbers for "particulate system" decreased, whereas all others increased (clearly, time intervals differ by a factor of 4, but the number of journals and the size of the research enterprise have increased also). It is then clear that this area, as indicated by the numbers above, has had a resurgence of interest. This was partly driven by physics and also by a smoother connection with fluid mechanics, fluid mechanics being a point of strength for the United States. The area is attracting small numbers of talented researchers, and there is an unmistakable trend upwards. However, research in this area is clearly driven by physicists. A survey of the key journals in this area, e.g., *Powder Technology* and *Granular Matter* (Tables 4.11 and 4.12) reveals that the U.S. contribution from 1997 to 2005 has declined from 27%-30% to around 20% for the first and has remained roughly constant at about 20% for the second. This is lower than the virtual congress assessment with 51% of the proposed speakers being from the United States, making it 8th from the bottom in U.S. participation in all areas surveyed. Chemical engineers have contributed about 25 to 30% of the publications in *Powder Technology* and 10%-15% in *Granular Matter*. Thus, the percentage of papers from U.S. chemical engineers was approximately 5%-6% for *Powder Technology* and about 3% for *Granular Matter*, both rather low.

Table 4.12 shows the trends in the two journals over the past 16 years. Although the number of U.S. papers has increased by 80% its relative percentage has been declining. Chemical engineers have more than doubled their contribution, but the relative percentage has remained the same. Asia (including Australia) dominates the numbers, but not the quality and impact (see Table 4.13), which is dominated by the European Union. Chemical engineers dominate the list of the 30 most cited, and U.S. chemical engineering has a good representation in this list at 30%-40%.

Relative Strengths and Weaknesses. There are several U.S. research centers concentrated around applications such as pharmaceuticals, fluidization, and energy. The picture here is almost the reverse of fluid mechanics. Since the 1960s fluid mechanics has been driven by fundamentals, and applications has followed. In particle technologies the situation has been exactly the opposite—the driver was applications, some of them very general, to be sure, but it is only recently that emphasis on more physics-like research focusing on general principles has reached chemical engineering.

In the area of characterization, industrial labs, in spite of the tremendous importance of solids processing across processes and products, have efforts that are manifestly less organized or standardized than labs focused

TABLE 4.11 Recent Contributions in *Powder Technology* and *Granular Matter*

	Powder Technology				Granular Matter		
	% U.S.	% EU	% China + India	% Chem. Eng.	% U.S.	% EU	% Chem. Eng.
1997	27	38	5	0			
2000	30	38	11	28	6	94	0
2003	15	49	16	29	25	75	14
2004	20	40	15	25	11	52	7
2005	21	42	15	0	21	69	10

TABLE 4.12 Publication Trends in *Powder Technology* and *Granular Matter*

	1990-1994		1995-1999		2000-2006	
		%		%		%
Total Number of Papers	749		863		1,783	
No. of U.S. Papers	209	28.00	211	24.00	360	20.00
No. of Chem. Eng. Papers	206	27.50	246	28.51	493	27.65
U.S., Chem. Eng.	47	22.82	76	30.89	123	24.95
EU, Chem. Eng.	20	9.71	32	13.01	77	15.62
Asia, Chem. Eng.	80	38.83	93	37.80	208	42.19
Canada, Chem. Eng.	18	8.74	17	6.91	29	5.88
S. America, Chem. Eng.	2	0.97	7	2.85	12	2.43

TABLE 4.13 Distribution of the 30 Most-Cited Papers in *Powder Technology* and *Granular Matter*

	1990-1994	1995-1999	2000-2006
No. of U.S. Papers	7	8	7
No. of Chem. Eng. Papers	17	20	17
No. of U.S. Chem. Eng. Papers	7	5	5
(% share among chemical engineering papers)	(41%)	(25%)	(30%)

on rheology, for example, in which there is a commonality of infrastructure across industries.

Opportunities for long-range funding in basic applications-free aspects of solids processing are virtually nonexistent in the United States.

Future Prospects. Historically the solid particles technology subarea has not been—in comparison with fluids—in the center of chemical engineering. A rejuvenation has taken place in the past decade, and significant progress has been made in problems ranging from the design and synthesis of "smart" particles to the ability to model extensive reaction schemes with complex fluid-particle hydrodynamics to the ability to study granular flows mixing and segregation using discrete-element methods. Much of this has been accompanied by advances in nonintrusive measurement techniques, such as magnetic resonance imaging, electrical capacitance volume tomography and positron emission for real-time solids flow measurements. Clearly all of these advances bring new science into chemical engineering.

Advances can be expected in drug delivery of particles (e.g., via inhalation), biomass (slurry) processing, and modeling of "real" particles (nonspherical, deformable, or cohesive, and particles with a wide size distribution), nanoparticle technology, and particle design for high pressure and high temperature for energy and environmental system applications.

Panel's Summary Assessment. The current U.S. position is "Among World Leaders," and in the near future, it will maintain this position.

4.2 AREA-2: ENGINEERING SCIENCE OF CHEMICAL PROCESSES

This area encompasses research in the science and engineering of processes, which are characterized, primarily, by chemical transformations. It has been divided into the following four subareas:

- catalysis
- kinetics and reaction engineering
- polymerization reaction engineering
- elecrochemical processes

4.2.a Catalysis

Catalysts accelerate the rate of chemical reactions by reducing their energy of activation. Catalysts also can improve the selectivity of chemical reactions by selectively catalyzing the rates of their different pathways. It has been said that over 95% of commercial chemicals and fuels are produced via catalytic reactions. Homogeneous catalysts are typically dissolved in the reaction medium, while heterogeneous catalysts are typically used in the solid phase.

The field of catalysis is core to chemical engineering. Relying on complex chemical and physical phenomena, catalysis interfaces with several disciplines including surface science, kinetics, solid-state materials science, and electrochemistry.

Catalysts are employed in large-scale industries, such as chemicals, hydrocarbon fuels, energy conversion, and transportation (e.g., automobile emission control). Catalysts are also employed in the manufacture of petrochemicals, specialty chemicals, pharmaceuticals, and polymeric materials. Correspondingly, progress in catalysis impacts the world's economy and well-being in significant positive ways.

U.S. Position. The Virtual World Congress in catalysis identified 144 speakers of whom 56% were from the United States when duplication in names was allowed and 50% when duplication was disallowed. The large non-U.S. representation is significant in light of the fact that 100% of those canvassed (experts) were from the United States.

Two of the leading catalysis journals are the *Journal of Catalysis* and *Applied Catalysis* (both A-General, and B-Environmental). Table 4.14 summarizes the publications data for these two journals. While the number of U.S. papers has increased from 1990-1994 to 2000-2006, the corresponding fraction of the total has decreased steadily from 33% (1990-1994) to 23% (1995-1999) and 15% (2000-2006). Chemical engineers have contributed 29%, 25%, and 21%, respectively, for the three periods examined. Of the chemical engineer authored papers, 60%, 50%, and 40% were U.S.-authored; notable is the stability of the European Union share (19%, 24%, and 22%) and the rapidly increasing Asian share (16%, 24%, and 36%).

We have also examined the citation statistics in the aforementioned suite of catalysis publications (Table 4.15). Among the 50 most-cited papers in these two journals, the share of U.S.-originated papers declined from 27 papers (i.e., 54% share) to 12 papers (i.e., 24% share); likewise the share

TABLE 4.14 Analysis of Publications in *Journal of Catalysis* and *Applied Catalysis*

	1990-1994	%	1995-1999	%	2000-2006	%
Total Number of Papers	2,255		3,932		6,859	
No. of U.S. Papers	747	33.00	891	23.00	1,047	15.00
No. of Chem. Eng. Papers	666	29.53	1,002	25.48	1,439	20.98
U.S., Chem. Eng.	401	60.21	496	49.50	576	40.03
EU, Chem. Eng.	127	19.07	238	23.75	323	22.45
Asia, Chem. Eng.	109	16.37	244	24.35	514	35.72
Canada, Chem. Eng.	31	4.65	35	3.49	59	4.10
S. America, Chem. Eng.	3	0.45	24	2.40	39	2.71
Internationalization (overlap)		0.75		3.49		5.01

TABLE 4.15 Distribution of the 50 Most-Cited Papers in *Journal of Catalysis* and *Applied Catalysis*

	1990-1994	1995-1999	2000-2006
No. of U.S. Papers	27	22	12
No. of Chem. Eng. Papers	27	28	14
No. of U.S. Chem. Eng. Papers	23	13	8
(% share among chemical engineering papers)	(85%)	(46%)	(57%)

of chemical engineering papers declined from 27% to 14%, while the share of U.S. chemical engineering citations shrank from 85% to 57%—which is still a leading position.

Although the United States represents about one-third of the world's economic activity, our role in catalysis falls below that fraction, and has been decreasing with time. The reduction of the U.S. role is being taken up by Asia, and it can now be projected that the number of Asian chemical engineering papers in catalysis will likely exceed the U.S. numbers in the coming 5-year period.

The general trend of decline in the relative share of U.S.-based participation in catalysis was also observed in our analysis of U.S.-, European Union-, and Asian-originated U.S. patents for the 5-year time periods discussed above.

Catalysis research is most often associated with large-scale chemical, petrochemical, or oil refinery processes. These business activities represent rapid growth areas in Asia while they are stagnant in the U.S. and in the European Union. It seems the Asian growth in catalysis R&D is mostly at the expense of a reduced U.S. share in overall R&D activity. It can be expected that the current (2006) upsurge in energy-related economic activity will reenergize U.S. R&D interest in catalysis, but it is unlikely that it will soon reverse the relative trends discussed above.

Relative Strengths and Weaknesses. While large-scale industrial catalysis originated in Europe (e.g., the Haber-Bosch synthesis of ammonia from hydrogen and nitrogen), modern catalytic science was, arguably, created in the United States, beginning with the evolution of the large-scale petroleum (e.g., fluid cracking catalysis) and petrochemicals (e.g., ethylene oxide) industry, during and immediately after World War II. This was followed by an upsurge in polymerization catalysis (e.g., polyolefins) then by emission control (e.g., automobile catalytic converters). These waves of catalytic technologies evolved on top of each other in a cumulative fashion, thus catapulting U.S. industrial catalysis R&D to world dominance and U.S. chemical engineering catalysis to a world-leadership position. However, the

Europeans caught up fast in petrochemicals and in emission control, followed by the Japanese. With the migration of large-scale catalytic process investments to either the source of oil (Middle East) or to the emerging markets (China, India), the rapid growth of the share in the world's catalysis research in Asia appears to be at the expense of research in the developed economies. While the relative share of U.S. catalysis research may be in decline, its absolute extent and its quality are not, as exemplified by still very strong academic efforts (e.g., 2005 Nobel Prize to two U.S. scientists working in catalysis, shared with one in Europe) and by the still leading position of U.S.-based authors in the most-cited papers.

Future Prospects. Among the most notable advances during the past 10 years are deep desulfurization catalysts for gasoline and diesel fuels; catalysts for olefin metathesis; extension of high-throughput and combinatorial techniques for catalyst development; asymmetric catalysts; solid acid-base catalysts; and computational chemistry for the design of homogeneous catalysts. Catalysis will remain the dominant technology in the world's petroleum, chemical, and polymer industries. It will also dominate both stationary and automobile emission control for a long while. Catalysis will also play an important role in the chemical conversion of biomass to chemical feedstocks and fuels, as well as the synthesis of new functional polymers for medical and biomedical applications and the electronics and communications industries.

The United States will remain a major source for scientific and technological progress, despite the rapidly growing competition in Asia. The current concerns about sufficient supply of petroleum and high energy prices will demand new, more energy-efficient and environment-friendly catalytic technologies. This is going to stimulate more R&D funding in catalysis, producing more graduates as well. Another major stimulus for catalysis research is the emerging and rapidly growing petrochemical industry in developing countries such as China and India. Catalyst and catalytic process research in these countries will take a significant part in generating new basic knowledge and commercially significant technologies. The Panel expects that the following will become problems of increased research interest: catalysis with chain shuttling; catalysts with higher surface area and supports with larger than 60% void space; partial oxidation of alkanes; direct fluorination of alkanes; methane activation catalysis; and coal conversion catalysis.

Catalysis, still being largely an experimental science, has been benefiting greatly from the introduction and rapid acceptance of combinatorial and parallelized high-throughput screening technologies. Equipment and software (to control the equipment and to analyze the data) are now available to conduct thousands of catalysis experiments per day, in an organized search for new and better catalysts or for improved products (such as metallocene catalysts for polyolefins and the resulting polymers themselves). It is expected that this massive acceleration of catalysis research will fuel rapid progress, especially when combined with traditional techniques. U.S. industrial research in heterogeneous catalysis is very strong and will continue to be so in the future. The European Union and Japan are two principal and nearly equal competitors. However, one should not overlook the progressive deterioration of the U.S. academic position in the area of heterogeneous catalysis, where a perception of having "peaked" exists. While the forecast for U.S. homogeneous catalysis research, driven primarily by chemists, is rather comforting, the forecast for academic heterogeneous catalysis is pessimistic.

Panel's Summary and Assessment. The current U.S. position is "Among World Leaders," and although in the future this position is expected to weaken, the United States will remain "Among World Leaders."

4.2.b Kinetics and Reaction Engineering

The subarea of chemical reaction engineering deals with the engineering aspects of chemically reacting systems. In the broader sense, these systems include the quantitative (usually model-based) analysis of chemical reactors of different types (batch or continuous; continuously stirred [CSTR] or plug-flow; fixed-bed, moving-bed, or fluidized; isothermal, non-isothermal, or adiabatic; one-phase or multiphase; catalytic or noncatalytic; homogeneous or heterogeneous, etc.). In a narrower sense, the core of the traditional chemical reaction engineering discipline is centered on the interactions of transport phenomena (heat, mass, and momentum transport) with chemical or catalytic kinetics in determining reactor behavior.

Reaction engineering covers a wide range of spatial scales, from meters (the scale of reactors), to centimeters or millimeters (the scale of catalysts), to microns and fractions of nanometers (the scale of catalyst pores), to fractional nanometers (the scale of catalytic surface phenomena).

Chemical reactors are the heart of most chemical processes, and thus reaction engineering is a core discipline in chemical engineering. Computational chemistry and molecular simulations, multiscale modeling, visualization of patterns in reacting systems, reactor design for large and complex reaction networks, multiphase reactors, microreactors, integrated bioreactors, and novel reactor configurations, are a few of the current research interests in kinetics and reaction engineering.

U.S. Position. *Chemical Engineering Science* is a leading journal for the field. The Panel examined 7,923 papers, published in three 5-year time intervals: 1990-1994, 1995-1999, and 2000-2006. While the field appears to be growing (number of papers increasing from 2,019 to 2,460 to 3,444), the U.S. share has been declining (30%, 23%, and 19%). European Union participation has been stable over this 15-year period (35%, 40%, and 38%), while Asian participation has grown (12%, 16%, and 22%). Notable is the European Union dominance in recent times (38%), followed by Asia (22%), and the United States (19%).

Of the eight experts for the Virtual World Congress, seven were from the United States. They proposed 142 speakers, 69% of whom were from the United States when duplication of names was allowed, or 50% when unique names were counted. The European Union was strongly represented; notable is the essential absence of proposed speakers from Asia. With chemical reaction engineering being a core area to chemical engineering, very few nonchemical engineers are active in the field.

Of the 100 most-cited papers, the US share has been declining (42%, 31%, and 23%), while the European Union share remained dominant (44%, 56%, and 51%). Asian papers were not so frequently cited 15 years ago (2%), but their share has increased significantly in recent years (12%).

The Virtual World Congress suggests that U.S. researchers lead the field, with Europe following as second. When we consider the volume of publications the European Union leads with United States being second and declining, and Asia third and growing. The European Union dominance in citations is also prominent.

Relative Strengths and Weaknesses. The origins of the field of chemical reaction engineering can be traced back to the 1940s and involve three essentially simultaneous papers by Thiele (United States), Damkoehler (Germany), and Zeldovich (Russia); they were the first to formally compute diffusion-reaction problems in porous catalyst pellets. The field blossomed first in the United States in the 1950s, and by the 1970s United States reaction engineering research represented the leading edge, exemplified by path-breaking mathematical modeling work for petroleum refining and automobile emission catalysis. The field had matured by the 1980s, and the United States lost its preeminence by the 1990s. U.S. chemical engineering departments have reduced their activity in reaction engineering, although the applications of the field are thriving in other subfields such as electronic and structural materials, polymers, biotechnology, and environmental science and technology. In the past 10 years the following developments have been among the most important: computation of kinetics for large and complex networks of chemical reactions; quantum-mechanical estimates of reaction rates; high-performance software for the analysis and design of complex, multiphase reactors; integrated microscale reactor configurations; and advances in reaction engineering of microelectronics fabrication.

The strengths of the U.S. position include a strong interdisciplinary approach built upon a broad and rigorous chemical engineering curriculum, and the broadening of the applications of reaction engineering principles to nontraditional fields, enumerated above.

The U.S. position has been weakened by the reduced industrial interest in large-scale process (catalysis) research and development, and the correspondingly reduced research funding in this area. Potentially encouraging is the recent resurgence of interest in energy-related chemical engineering problems, and it is expected that chemical reaction engineering thinking and problem-solving approaches (e.g., for fuel cells) will make significant contributions.

Future Prospects. The field of reaction engineering might be revitalized during the coming decade due to a combination of increasing industrial needs (pressures of energy supply and pricing, global competition, local and global environmental regulations) and increasingly sophisticated and powerful computational capabilities (e.g., computational fluid mechanics, computational chemistry, and new analytical techniques). There is also a trend to broaden the field to include stochastic model-building techniques beyond the traditional, deterministic models; examples include reactor models based on neural networks, data mining and filtering, adaptive control, and Monte Carlo techniques. Yet another trend is to embrace the (larger) field of combustion in reaction engineering and expand the field to include reaction engineering at the microscale. The interface between chemical reaction engineering and biological reaction engineering has been tenuous, and techniques and methodologies developed within the scope of the former have not found their full way to applications within the scope of the latter.

Panel's Summary Assessment. The current U.S. position is "Among World Leaders," and although in the future this position is expected to weaken, the United States will remain "Among World Leaders."

4.2.c Polymerization Reaction Engineering

Development of synthetic polymers has been one of the most successful achievements of the chemical industry in the past 80 years, with numerous applications in the fiber, rubber, plastics, and coatings industries. In the early years, polymers like polystyrene or high-pressure polyethylene were produced mainly by radical polymerization, but catalytic polymerizations attained rapid industrial importance. Two very important innovations have been polycondensation reactions for the production of polyamides and polyesters by DuPont, in the United States, and the Ziegler-Natta catalytic polymerization for the production of isotactic polypropylene and other polyolefins, initiated in Europe in the late 1940s. More recently, catalytic systems have become sophisticated using ionic catalysts, metallocenes, and other single-site catalysts. Radical polymerization control methods have also improved and use different systems of living radicals or transfer agents. UV-induced polymerizations have become particularly important in sound replication (i.e., systems for storage and replication of sound and information, such as CDs, DVDs) and biomedical polymers.

The recent evolution in polymerization technologies has allowed a fine control and monitoring of the microstructure of polymers produced, while achieving the usual requirements for an industrial chemical process: safety, environmental concerns, and productivity. Eventually, engineers in this field must deliberately and accurately control the interaction among polymerization conditions, the resulting polymer microstructure and the polymer properties. Control of the polymer microstructure is related to the molecular weight distribution, stereoregularity/stereotacticity, copolymerization conditions, grafting, etc. Advanced innovations in this field have led to multibloc polymerizations, giving access to improved mechanical properties due to control of the microstructure of the polymeric materials.

U.S. Position. Eleven experts (10 from the United States) suggested 89 scientists and engineers for the Virtual World Congress with 44% of them from the United States, indicating a leading U.S. position in this subarea.

This is an interdisciplinary area with chemical engineers contributing between 15% (*Macromolecules* and *Journal of Polymer Science Part A: Polymer Chemistry*) and 50% (mainstream chemical engineering journals: *I&EC Research, AIChE Journal, Chemical Engineering Science*) of the total publications with the balance contributed primarily by chemists and material scientists. There is no specialized journal for publication of polymerization reaction engineering studies. Most publications in the field are published in *Macromolecules* and the *Journal of Polymer Science Part A: Polymer Chemistry* (if related to kinetics), *Industrial Engineering Chemistry Research, Chemical Engineering Science,* and *AIChE Journal* (if of an engineering and modeling nature) and the *Journal of Polymer Engineering* (if of an applied nature). From Table 4.16 we see that in the area of polymer synthesis and kinetics U.S. contributions have more than doubled from the 1990-1994 to the 2000-2006 period, but their percentage of the total has been reduced from 40% to 30%. We also notice a similar reduction (from 48 to 29%) in the corresponding percentage of U.S. contributions in the area of polymerization modeling, engineering, and control (see Table 4.17) despite an almost 3-fold increase in the number of papers. Both of these

TABLE 4.16 Distribution of Polymerization Reaction Engineering Published and Most-Cited Papers in *Macromolecules* and *Journal of Polymer Science: Polymer Chemistry Part A*

	1990-1994		1995-1999		2000-2006	
		%		%		%
Total Number of Papers	1,873		2,991		5,701	
No. of U.S. Papers	753	40.20	1,001	33.47	1,744	30.59
No. of Chem. Eng. Papers	195	10.41	342	11.43	777	13.63
U.S., Chem. Eng.	97	49.74	195	57.02	334	42.99
EU, Chem. Eng.	22	11.28	31	9.06	103	13.26
Asia, Chem. Eng.	68	34.87	119	34.80	345	44.40
Distribution of 30 Most-Cited Papers						
U.S.	13		12		9	
EU	6		7		11	
Asia	5		10		9	
Canada	6		1		1	

TABLE 4.17 Distribution of Polymerization Reaction Engineering Published and Most-Cited Papers in *I&EC Research*, *AIChE Journal*, and *Chemical Engineering Science*

	1990-1994		1995-1999		2000-2006	
		%		%		%
Total Number of Papers	112		247		491	
No. of U.S. Papers	54	48.21	103	41.70	144	29.33
No. of Chem. Eng. Papers	69	61.61	158	63.97	269	54.79
U.S., Chem. Eng.	47	68.12	79	50.00	100	37.17
EU, Chem. Eng.	7	10.14	20	12.66	57	21.19
Asia, Chem. Eng.	9	13.04	31	19.62	73	27.14
Distribution of 30 Most-Cited Papers						
U.S.	17		13		12	
EU	5		8		8	
Asia	3		2		5	
Canada	4		6		4	
S. America	1		0		0	
Other	0		0		1	

trends are due to much faster (5-fold to 8-fold) growth in the number of publications from European Union and Asian countries. In terms of quality and impact the distribution of most-cited papers indicates that European Union and Asian countries have established a parity with U.S. contributions in the area of synthesis and kinetics (see the most-cited papers list for 2000-2006 in Table 4.16), but the U.S. publications maintain a dominant position on problems related to the engineering of polymerization processes (Table 4.17).

In addition to academic research, industrial R&D by the U.S. petrochemical companies has played a significant role in the development of innovative polymerization processes (e.g., metallocenes, single-site catalysis). For example, analysis of the patents on polymerization catalysts filed in the United States during the period 1990-2005 indicates that U.S. companies filed as many patents (about 1,700) as European Union (about 900) and Japanese (about 800) companies combined. Such a strong technological position has allowed U.S. companies to be world leaders in licensing polymerization processes around the world.

In conclusion, the leadership of U.S. research in polymerization reaction engineering remains strong.

Relative Strengths and Weaknesses. The subarea of polymerization reaction engineering is interdisciplinary in nature as it involves chemists, chemical engineers, and practicing polymer engineers. The contributions of chemical engineers have been pivotal, as they provide much needed understanding of advanced reaction engineering methods for better design and control of polymerization reactors. Polymerization reactors come in a wide range of systems from gaseous to homogeneous noncatalytic or catalytic reactors, as well as bulk, solution, suspension, and emulsion polymerization reactors. Significant stability and control problems are faced due to high operating viscosities and the associated autoacceleration and other effects. For these reasons U.S. and European chemical engineers who were educated early on in reaction engineering principles have been major contributors to this field.

The close cooperation of academic and industrial teams through research consortia has been a significant strength of U.S. research in polymerization reaction engineering. Furthermore, the strong safety-oriented culture of the U.S. chemical industry has expanded the scope of polymerization reaction engineering to include aspects of process systems engineering, such as dynamic modeling and operational scheduling and control, and has led to a significant differentiation of U.S. research activities in this subarea. However, as the strategies of the U.S. and European chemical companies focus more and more on the Asian polymers market, it is reasonable to expect that the research center may migrate to Asia as well. Indeed, the number of Asian publications and their quality in polymerization reaction engineering have steadily improved, especially in the past 10 years (see Tables 4.16 and 4.17).

Future Prospects. Metallocene and postmetallocene catalysis have been among the most significant advances in polymerization reaction engineering during the past 10 years. In addition, controlled living free-radical polymerization, atom-transfer radical polymerization, polymerization in supercritical CO_2, and dendrimer polymerization have had significant effects. Increased focus on the interplay between polymerization operations and resulting micro- and macrostructure and properties of polymer products will require creative new approaches for polymerization reaction engineering. The United States has strong position in this area to address the needs for the production of a variety of new products; cost-competitive block copolymers, conducting and semiconducting polymers, self-assembled polymers during polymerization, polymers with dynamic response, self-healing polymers, polymers from biomass, polymers from bacteria and plants, and others. The European Union and Japan are very strong in this area while the other Asian countries are making significant strides in more classical technologies.

Panel's Summary Assessment. The current U.S. position is "Among World Leaders," and in the future, the United States will maintain this position.

4.2.d Electrochemical Processes

Electrochemistry and electrochemical engineering have far-reaching technological significance. Electrochemical processes for the manufacture of chemicals (e.g., chlorine) and metals (e.g., aluminum), electrochemical storage batteries (e.g., lead-acid for cars, lithium for computers and cell phones), electroplating (e.g., in microelectronic chip and circuit board manufacture or for structural and decorative purposes in the car industry), electroorganic synthesis of chemicals, electrochemical sensors (e.g., glucose sensor in blood), fuel cells, and many other applications are the result of our knowledge about and ability to manipulate electrochemical processes.

Despite the significance of the field, it is no longer viewed as a core area in academic chemical engineering departments. Very few concentrated graduate-level educational and research efforts in electrochemical engineering still exist within chemical engineering departments in the United States; among these few are Berkeley, Case Western, and the University of South Carolina.

Electrochemical engineering overlaps with several fields, such as catalysis, materials science, and biomedical engineering, and thus practitioners of the discipline may not be easily identified as chemical engineers.

U.S. Position. The Virtual World Congress poll yielded 67 speakers, 57% of whom are from the United States when name duplication is allowed or 50% when duplications are disallowed. Five journals were analyzed for U.S. representation in the electrochemical engineering literature. Three of these, *Journal of the Electrochemical Society*, *Journal of Applied Electrochemistry*, and *Electrochimica Acta*, were grouped together for this analysis (Table 4.18). For the three periods analyzed, 1990-1994, 1995-1999, and 2000-2006, the total number of papers increased from 6,305 to 7,017 to 10,089; U.S. papers represented a decreasing set of percentages, from 34% to 28% to 23%. Interestingly, the percentage of chemical engineering papers (of the total number of papers published) held constant at the level of 11%-12%.

The chemical engineering papers were further analyzed for their geographical origins; the United States had a dominating, although declining, share: 63%, 57%, and 44%, respectively. The European Union share (8%, 13%, and 13%) appears to have been modest, while a strong and growing Asian publishing activity is evident (23%, 27%, and 43%). Of the 50 most-cited papers in these journals, few were written by chemical engineers.

We have also analyzed the publications in two more specialized electrochemical journals, *Journal of Power Sources* and *Solid State Ionics*. In the former, 5%-13% of papers were from chemical engineers; in the latter, 2%-5%. However, in both cases, publication by chemical engineers has been increasing with time with the largest (and growing) component being from Asia.

TABLE 4.18 Analysis of Publications in *Journal of Electrochemical Society*, *Journal of Applied Electrochemistry*, and *Electrochimica Acta*

	1990-1994		1995-1999		2000-2006	
		%		%		%
Total Number of Papers	6,305		7,017		10,089	
Total No. of U.S. Papers	2,165	34	1,982	28	2,336	23
Total No. of Chem. Eng. Papers	694	11.01	746	10.63	1,299	12.88
U.S., Chem. Eng.	435	62.68	427	57.24	570	43.88
EU, Chem. Eng.	53	7.64	96	12.87	172	13.24
Asia, Chem. Eng.	157	22.62	205	27.48	554	42.65
Canada, Chem. Eng.	46	6.63	24	3.22	46	3.54
S. America, Chem. Eng.	6	0.86	3	0.40	14	1.08
Internationalization (overlap)		0.43		1.21		4.39

Relative Strengths and Weaknesses. The data collected from the Virtual World Congress poll and the literature review indicate that the strength of U.S. electrochemistry and electrochemical engineering is approximately proportional to the U.S. share of publications in the world economy. The U.S. share is declining, the European Union share is stable, and the Asian share is increasing.

U.S. strengths include a strong and broad education of chemical engineers, able to deal in depth with a wide variety of technical challenges that involve, but are not restricted to, electrochemical problems. Another U.S. strength is the thriving venture-capital system, which allows small, technology-oriented companies to be spawned from breakthrough university, or government-funded research programs; notable examples are fuel cells, advanced batteries, and biomedical sensors. A possible U.S. weakness is the rapid drift of electrochemical engineering away from the core educational curriculum.

Among the most notable developments during the past 10 years are the following: advances in rechargeable lithium ion batteries with liquid or polymer electrolytes; advances in fuel cells with proton-conducting membranes; electrochemical sensors for blood glucose level monitoring; and room-temperature solid electrolytes.

Future Prospects. Electrochemical engineering is gaining increased relevance again, due in part to the world's repeated energy crises. Electrochemical engineering could be key to a future hydrogen economy (e.g., nuclear energy–powered water electrolysis, fuel cells), and with a possibly abundant future electrical energy supply, it could be the basis of an increasing share of electroorganic synthetic processes in the chemical industries. In the shorter run, there is a huge demand for portable and mobile electrical energy storage (laptops, personal communicating devices, hybrid electric vehicles), and this will likely stimulate academic R&D involvement and generate R&D support (e.g., the Department of Energy lithium battery program managed at the Lawrence Berkeley National Laboratory). However, the same events drive electrochemical R&D in other parts of the world and at increasingly higher levels. So, the relative U.S. position may not be as strongly affected as the magnitude of the U.S. effort. Japan is the major competitor in this subarea with significant industrial and government investments directed towards R&D of new materials for batteries and fuel cells. Other Asian countries such as Korea have been making significant investments in this subarea.

Panel's Summary Assessment. The current U.S. position is "Among World Leaders," and although in the future this position is expected to weaken, the United States will remain "Among World Leaders."

4.3 AREA-3: ENGINEERING SCIENCE OF BIOLOGICAL PROCESSES

This area encompasses research on the science and engineering of processes, which are characterized, primarily, by biological transformations. It has been divided into the following four subareas:

- biocatalysis and protein engineering
- cellular and metabolic engineering
- bioprocess engineering
- systems, computational, and synthetic biology

Biomedical products and biomaterials are considered separately in Area-6 (see Section 4.6).

Chemical engineering is an important contributor in each of the four subareas, which are also influenced by significant contributions from other disciplines, notably chemistry and/or biology. In general (as evidenced mainly by the journal analysis), within each subarea U.S. chemical engineers are playing a major role, or are sharing a prominent role with nonchemical engineers. The influence of chemical engineering appears to have been, and continues to be, strongest in bioprocess engineering and in cellular/metabolic engineering, where chemical engineers have played a leading role. Nonetheless, chemical engineering has yet to realize the full potential of its continuing and progressive integration of biology and its unique approach to the analysis, synthesis, and design of complex systems on multiple scales.

It should be noted that these subareas are fairly young in comparison to the more traditional subareas of chemical engineering, and the history of chemical engineers working in them is relatively short. Furthermore, the journal analysis does not accurately reflect the many fundamental and practical innovations emerging from industry, e.g., the pharmaceutical and biotechnology industries, which are intellectually fertile industries and in the latter case at least, is clearly a sector of U.S. dominance. For example, the steady growth of commercial biotherapeutics is testimony of efficient production methods and manufacturing technologies that utilize cellular engineering and bioprocess engineering, among many other engineering contributions. Furthermore, as this industry continues to mature and bring more products to market, chemical engineering can be expected to play an ever-increasing role. However, the extent to which such activity will take place in the United States versus overseas, as manufacturing facilities expand abroad, is an open question.

U.S. Position. In each subarea, the contribution of biology is very significant, even in journals with very strong engineering orientation. For example, the percentage of papers from biologists in the journal *Biotechnology & Bioengineering* increased from 12% in 2000 to 23% in 2005 (Table 4.19). The percentage from chemical engineers decreased from 1990-1994 to 2000-2006, but has remained roughly constant during the past 5 years (31% ± 2%, based on data from 2000, 2003, 2004, and 2005; see Table 4.21). Furthermore, as Table 4.20 shows, contributions from chemical engineers to the journal *Biotechnology Progress* decreased from 56% (1990-1994) to 37% (2000-2006), and from 36% in 2000 to 28% in 2005. Although these numbers suggest a declining influence of chemical engineers, Table 4.22 indicates that the impact of papers published by U.S. chemical engineers is exceptionally strong. In addition, of the top 100 papers published from 2000-2006 in *Biotechnology & Bioengineering*, U.S. chemical engineers contributed 23%, and the corresponding percentage for *Biotechnology Progress* over the same period is 31%.

4.3.a Biocatalysis and Protein Engineering

Biocatalysis, and to a lesser extent the newer field of protein engineering, has long been an area of active international participation. Biocatalysis encompasses the use of enzymes and whole cells to carry out biotransformations on a wide range of scales, from analytical devices to industrial processes. Enzymes are often used in heterogeneous formulations, which raise many of the same issues as heterogeneous chemical catalysts for reaction engineering and reactor design. Immobilized enzymes and whole cells are examples of heterogeneous biocatalysts, which were initially developed for practical applications primarily in Europe and Japan, respectively. Enzymes

TABLE 4.19 Analysis of Publications in *Biotechnology & Bioengineering*

	1990-1994		1995-1999		2000-2006	
		%		%		%
Total Number of Papers	1,323		1,605		2,275	
Total No. of Chem. Eng. Papers	568	42.93	547	34.08	737	32.40
U.S., Chem. Eng.	400	70.42	342	62.52	406	55.09
EU, Chem. Eng.	31	5.46	69	12.61	134	18.18
Asia, Chem. Eng.	65	11.44	66	12.07	144	19.54
Canada, Chem. Eng.	50	8.80	42	7.68	51	6.92
S. America, Chem. Eng.	0	0.00	6	1.10	23	3.12
Internationalization (overlap)		39.06		30.06		35.24

BENCHMARKING RESULTS: BY AREA OF CHEMICAL ENGINEERING

TABLE 4.20 Analysis of Publications in *Biotechnology Progress*

	1990-1994		1995-1999		2000-2006	
		%		%		%
Total Number of Papers	366		570		1,368	
Total No. of U.S. Papers	274	75	302	53	541	40
Total No. of Chem. Eng. Papers	204	55.74	253	44.39	513	37.50
U.S., Chem. Eng.	167	81.86	181	71.54	286	55.75
EU, Chem. Eng.	24	11.76	14	5.53	65	12.67
Asia, Chem. Eng.	16	7.84	49	19.37	133	25.93
Canada, Chem. Eng.	7	3.43	9	3.56	26	5.07
S. America, Chem. Eng.	1	0.49	3	1.19	17	3.31
Internationalization (overlap)		5.39		1.19		2.73

TABLE 4.21 Percentage of Papers Published in *Biotechnology & Bioengineering* That Include an Author with a Chemical or Biochemical Engineering Affiliation

	Based on Total Papers (all chemical engineers)	Based on Most-Cited Papers (U.S. chemical engineers)
2003[a]	29	20
2004[a]	34	34
2005[b]	29	24

[a]Based on top 50 papers (with most citations).
[b]Based on top 42 papers (with 3 or more citations).

TABLE 4.22 Distribution of the 30 Most-Cited Papers

	Biotechnology & Bioengineering			Biotechnology Progress		
	1990-1994	1995-1999	2000-2006	1990-1994	1995-1999	2000-2006
Total No. of U.S. Papers	19	9	14	23	23	21
No. of Papers Chem Eng.	21	18	18	20	20	18
No. of Papers U.S. Chem Eng.	14	6	10	14	16	13
(% share among chemical engineering papers)	(70%)	(33%)	(55%)	(70%)	(80%)	(70%)

suspended in nonaqueous media are another type of heterogeneous biocatalyst, an innovation with principal roots in the United States.

Protein engineering, a relatively recent breakthrough advance for the design and application of proteins, comprises various genetic techniques that enable the development of new enzymes with improved properties (including the potential for biotherapeutic applications) and infuses elements of protein chemistry and molecular biology into the domain of industrial biocatalysis. This emerging field is an outgrowth of genetic engineering, whose origins are centered in the United States. In addition, the prediction and simulation of protein structure and function are research topics that continue to attract and benefit from the participation of chemical engineers.

The Virtual World Congress proposed for this subarea includes 130 nominations, 54% of which were for U.S.-based researchers, the lowest percentage of the four subareas. When duplication of names was disallowed, the percentage of U.S. participation dropped to 42%. The leading journals in this area include *Angewandte Chemie International*, which is primarily a chemistry journal, and *Biotechnology & Bioengineering*, which is primarily an engineering journal. In addition to the two already mentioned, the following three journals were included in the analysis of publications and citations: *Proteins: Structure, Function, and Bioinformatics*, *Protein Science*, and *Enzyme and Microbial Technology*. In 2005 U.S. authors contributed 34% of the papers on average. However, on average, only 11% of the total contributions in 2005 were from chemical engineers. Furthermore, two of these five journals (*Biotechnology & Bioengineering* and *Enzyme and Microbial Technology*) accounted for 84% of the contributions from chemical engineers. These percentages reflect the international and interdisciplinary nature of this subarea, and indicate that U.S. chemical engineering occupies a significant, but clearly nondominant position.

4.3.b Cellular and Metabolic Engineering

This subarea is of growing importance within chemical engineering and combines elements of cellular and molecular biology with reaction engineering and control theory. Rational strategies for manipulating the behavior of cells for functions ranging from protein or small molecules production to more favorable growth characteristics are proving useful in the advancement of biotechnology for many applications, and hold great promise as an instrument of future breakthroughs throughout the industry. Metabolic engineering has evolved into a codified discipline largely through the vision and efforts of U.S. chemical engineers. Furthermore, chemical engineers are well suited to lead this field because of their facility in analyzing and optimizing large reaction networks and their understanding of feedback regulation mechanisms in complex interactive systems.

Of the leading journals in this field, three are engineering oriented (*Metabolic Engineering, Biotechnology & Bioengineering,* and *Biotechnology Progress*), and one is in applied microbiology (*Applied and Environmental Microbiology*). In 2005 U.S. authors contributed 42% on average of the papers published in these four journals, with chemical engineers contributing only 20%. The proportion of U.S. speakers among the Virtual World Congress participants was 75% (independent of allowing or disallowing name duplications) with 57 unique U.S. speakers. In all, these percentages reflect a strong U.S. position in this subarea, but the percentage contribution of U.S. chemical engineering research was less than expected, especially in comparison to biology (which was 23%).

4.3.c Bioprocess Engineering

Biochemical process engineering is arguably the most well-established biorelated subarea of chemical engineering. This area has seen steady growth since the phenomenally successful scale-up of antibiotics production that began in the 1940s. The advent of recombinant DNA technology in the 1970s and the rapid emergence of mammalian cell culture were two further developments that spurred great interest and progress in bioprocess engineering, and firmly established it as a critical discipline within chemical engineering. Biochemical processes now encompass a diversity of biological systems, and include technologies over a wide range of scales, from the nanoscale to production scales. The rise of biotechnology products within the marketplace for pharmaceuticals (e.g., the glycoprotein erythropoietin, which is manufactured and marketed by U.S.-based companies, is now among the 10 top-selling drugs in the world) is but one example of how recent developments in chemical and bioprocess engineering are making a major impact.

In 2005 U.S. authors contributed 23% on average of the papers published in the leading four journals, with chemical engineers contributing 24% and biologists 18%. Of the Virtual World Congress participants in this subarea, 63% (duplications disallowed) or 69% (duplications allowed) were from the United States.

4.3.d Systems, Computational, and Synthetic Biology

This subarea actually encompasses separate but overlapping subareas, all of which interface closely with biological sciences. It is a relatively new (and rapidly evolving) area within chemical engineering that caters to the quantitative mindset and aptitudes of chemical engineers, their facility with computational methods, and their ability to analyze complex and interactive reaction networks. It is thus recognized as a ripe area for growth within chemical engineering and one that chemical engineers would appear well suited to lead. Synthetic biology parallels metabolic engineering in its objective of developing cellular systems with improved synthetic capabilities, but is broader in that it goes beyond metabolic pathways to encompass even more complex aspects of cellular and organismal functions.

Of the journals considered for this field, four are engineering oriented (*Biotechnology & Bioengineering*, *Metabolic Engineering*, *Biotechnology Progress*, and *Computers and Chemical Engineering*, whose scope extends well beyond biological systems), and one is *Bioinformatics*. In 2005 U.S. authors contributed 41% on average of the papers published in these four journals, with chemical engineers contributing 26% (however, the contribution to the journal *Bioinformatics* was only 1%). Notably, 78%-79% of the Virtual World Congress participants in this subarea were from the United States, the highest of all four subareas.

Relative Strengths and Weaknesses. Relative strengths and weaknesses of chemical and biochemical engineering (broadly defined here to encompass the four subareas discussed above) are summarized below.

The interdisciplinary outlook of chemical engineering and its integrative approach to the analysis, synthesis, and design of complex systems provides an ideal intellectual platform to interface with and complement biology. In this regard, chemical engineering is poised to play a leading role in the expanding quantification of biology, and has been at the forefront of such efforts for many years.

In addition, the chemical engineering curricula at many universities have been expanding to include greater emphasis on the biological sciences and biotechnology, and in many cases department names, faculty and student distributions, and degree options are changing to reflect this emphasis. The expanding portfolio of biotechnology products and the associated manufacturing needs will present a steady demand for chemical engineering expertise in the marketplace. Continuing advances and dynamism in biology will provide abundant opportunities for chemical engineers to facilitate the advent of new products and processes; likewise, as chemical engineering encompasses more biology, it will make greater contributions to the discovery of new information in biological fields.

Growing competition from overseas production facilities may undercut opportunities in the manufacturing sector for U.S. biochemical engineers, especially at the bachelor's level.

Funding for biorelated chemical engineering research has not kept pace with the expanding scope and popularity of the field. In general, funding for biorelated research is dominated by National Institutes of Health funding for medically-oriented projects. Bioprocess engineering will contract due to funding limitations. The field is also at risk of becoming overshadowed by biomedical engineering because of the imbalance in resources.

As biochemical engineering becomes more diverse and specialized, there is a growing risk that it will become separate and distinct from the rest of chemical engineering and/or suffer destabilizing fragmentation within itself. As a field, chemical engineering must develop a central academic curriculum and professional identity that encompasses its newer and emerging subfields, including biochemical engineering, while preserving unifying and distinguishing themes.

Future Prospects. The following developments have been among the most notable advances in the past 10 years: laboratory evolution and rational design strategies for the creation of enzymes and antibodies with novel functions; design of peptides with antimicrobial activity; de novo design of peptides and proteins; bacterial and yeast systems for surface display of polypeptides with improved biological functions and biotechnological properties; incorporating metabolic engineering into practice within the pharmaceutical industry; microarray technologies and microscale platforms for genetic and biochemical analyses, including human toxicology; development of herbicide-tolerant and insect-resistant biotech crops; and sequencing of human, plant, and animal genomes.

Biotechnology will continue to evolve and expand as an underpinning technology of U.S. competitiveness in academia as well as industry. As fundamental understanding of biology continues to expand, from the molecular to the organismal level, so will opportunities for chemical engineering to enhance and exploit that understanding. Such opportunities will include more traditional avenues such as manufacturing processes for biotherapeutics and biofuels, as well as emerging challenges associated with increasing engineering capabilities for biomolecular and cellular design and the implementation of design principles within the paradigm of synthetic biology. From improved health care to renewable energy to a clean and sound environment, biochemical engineering, as defined by the sum of the four subareas, has enormous potential for positive impact. In this regard, as chemical engineering shifts or expands its focus more toward product design, as opposed to process design, biological systems could occupy a position of even greater importance. With regard to promising funding initiatives, the Genome to Life program and the Department of Energy's initiative on biofuels offer significant opportunities for chemical engineering research.

Maintaining a position of prominence in this area is of critical importance. The competitive position of U.S. chemical engineering in this area is strong, with chemical engineering having played a major role in the inception and development of entire fields. However, signs point to a situation where this position is not improving, and is possibly declining in relation to expansion of other disciplines. This is particularly true in biocatalysis and the systems area, both of which bridge the traditional and modern paradigms of chemical engineering and are ripe for innovation and further development (as are all of the subareas considered). These areas, at least, require strengthening to avoid further slippage against an expanding base of international competition.

The Panel expects that the following subjects will attract researchers' attention in the future:

• **Biocatalysis and protein engineering:** biomimetic catalysts and functionally robust enzyme mimics; biocatalytic composites combining enzymatic function and advanced material properties including nanoscale structures; advanced algorithms for protein structure-function prediction
• **Cellular and metabolic engineering:** integration of stem cell biology into tissue engineering; design of cellular systems with specialized functions and/or enhanced synthetic capabilities
• **Bioprocess engineering:** efficient and sustainable production of biofuels; plant-based vaccines and biotherapeutics; cell-free systems for the synthesis and production of bioproducts; marine biotechnology and marine-derived processes
• **Systems, computational, and synthetic biology:** integration of modeling, analysis, and design of genetic and metabolic processes; elucidating the structure of signal transduction pathways; creation of novel biological entities and technologies through the assembly of disparate "parts" from multiple sources

Panel's Summary Assessment

• **Biocatalysis and protein engineering:** The current position is "Among World Leaders," and in the future the United States will be "Gaining or Extending" this position relative to others.
• **Cellular and metabolic engineering:** The current position is at the "Forefront," and in the future U.S. research in this subarea will be "Gaining or Extending" this position relative to others.
• **Bioprocess engineering:** The current position is "Among World Leaders," and in the future the United States will be "Maintaining" this position relative to others.
• **Systems, computational and synthetic biology:** The current position is at the "Forefront," and in the future the United States will be "Maintaining" this position relative to others.

4.4 AREA-4: MOLECULAR AND INTERFACIAL SCIENCE AND ENGINEERING

Molecular and interfacial science and engineering refers to the study of the structure, properties, and fabrication of large molecules (usually organic molecules or polymers) and molecular aggregates, and of the interfaces between different phases and/or materials. The relevant length scale is generally between a fraction of a nanometer and roughly 100 nanometers. As a result of this governing length scale, it is not surprising that there is significant overlap between this area and nanostructured materials, discussed in Section 4.5.d of this report.

One of the distinguishing features in this subfield is the importance of the molecule itself: from its chemical composition and conformation to the way in which an ensemble of molecules assembles into a larger entity such as a two-dimensional film, a single three-dimensional nanostructure, or a topologically complex larger structure. Such different structures can in some cases be achieved in a single system simply by varying the conditions under which the structure is formed. Potential end uses are extremely varied and include applications such as purification, filtration, molecular recognition and sensors, drug delivery vehicles, molecular electronic devices, structured materials, and others. The youth of the field suggests that many additional applications will be identified and realized over time.

The emphasis on control and exploitation of molecular structure has meant that the field has been substantially populated by chemists, chemical engineers, and materials scientists. All of these fields are heavily represented in the data examined in the preparation of this report. In some cases, it has been possible to separate the contributions by chemical engineers; in other cases it has proven difficult or impossible.

U.S. Position. Polling of 15 world leaders in molecular and interfacial science and engineering resulted in the identification of 166 unique Virtual World Congress speakers, 63% of whom were U.S. based. When multiple nominations of the same individual are allowed, 69% of the 168 nonunique nominations were U.S. based (see Table 3.1). Analysis of the names of the speakers shows that the European Union made up most of the non-U.S. nominations, with notable contributions from Israel, Canada, and Asia.

The results of the Virtual World Congress poll show the United States to be in a "Dominant, at the Forefront" position today (60%-70%), followed by the European Union. Despite the dominance of materials scientists and chemists in this field, chemical engineers have significant presence.

Analysis of publications in the top journals of molecular and interfacial science and engineering shows general consistency, with some interesting differences from the above results. The top journals for the field are *Langmuir*,

Journal of Physical Chemistry B, Journal of Colloid and Interface Science, Macromolecules, and *Colloids and Surfaces A.* The total number of papers in all of these journals in 2005 was 8,088. Of these, 32% had authors from the United States, 39% from the European Union, 11% from China, and 4% from India. In addition, 13% had authors affiliated with a chemical engineering department. Although the total number of papers published in these journals in 2004 was 20% less than in 2005, the percentages of U.S.-based, India-based, and chemical engineering-department affiliated papers were unchanged. European Union-based authors accounted for 2% more papers than in 2005, while China-based authors accounted for 3% (less than a year later). Earlier data were available only for *Journal of Colloid and Interface Science, Colloids and Surfaces A, and Macromolecules.* Table 4.23 shows the results for these three journals over the time period for which data are available.

Relative Strengths and Weaknesses. The Virtual World Congress analysis shows that today the United States is a world leader in this field with 60%-70% of the top experts identified in the exercise. In addition, U.S. leaders were more likely to be chosen by multiple organizers, indicating that they were particularly widely recognized. It is important to recognize, however, that these results could be somewhat skewed by the preponderance of U.S.-based respondents to the Virtual World Congress survey.

Publication analysis shows that for the journals studied, the field is very global in nature. U.S. and European Union contributors account for the majority of papers, with the European Union as the single leading region over the period examined. The dominance of these two regions has diminished over the past decade, however, falling from nearly 75% in 1997 to 62% in 2005. The rise in the percentage of publications from China accounts for this fall.

TABLE 4.23 Papers Published in *Journal of Colloid and Interface Science, Colloids and Surfaces A, and Macromolecules* for 3 Years over the Past Decade

	1997	2000	2005
Total Number of Papers	2,002	2,400	3,362
No. of U.S. Papers	665	705	875
%, U.S.	33	29	26
%, EU	40	40	36
%, China and India	6	7	22

The combination of these two findings suggests that the intellectual leadership in this field has a significant U.S. component. At the same time, total U.S. participation is well under half of the total activity in the field, and the fraction of work carried out in the United States is shrinking. The United States continues to achieve rough parity with the European Union. Significant caution about the future is warranted due to the rapid rise of the rate of Chinese publications at the expense of both the United States and the European Union. Further anecdotal support for this interpretation is provided by the list of the "25 hottest papers" in *Colloids and Surfaces A* for the period January-March 2006. Only five of these papers had U.S. authors (three of these with authors from at least one other country). Eight papers had European Union authors, and nine had authors from China.

Future Prospects. Molecular and interfacial science and engineering represents one of the primary ways that chemical engineers interact with the enormous challenges and opportunities presented by nanotechnology. The National Nanotechnology Initiative has provided substantial funding to academic chemical engineering researchers to advance in many areas, very often in close collaboration with researchers from other disciplines. This research also connects directly to problems in thermodynamics and rheology, and often leads to the synthesis of new materials. Examples are improvements in understanding the mechanisms of friction and wear (tribology), polymerization in confined geometries, development of remarkably efficient new sensors, and generation of a variety of sophisticated microfluidics devices. Equally impressive are the advances in theory and simulation that have illuminated our understanding of block copolymer self-assembly, surfactant micellization, polymer/colloid interactions, and more fundamental issues such as the hydrophobic effect. Chemical engineers are also well placed to apply the x-ray and neutron characterization tools now available at synchrotron and the new advanced spallation neutron sources.

The United States is currently among the world leaders in molecular and interfacial science and engineering. The current major competitor is the European Union; the United Kingdom, France, and the Netherlands show particular strength. Asia is not a significant player today, although there are signs that its position may be evolving quickly. The U.S. standing is roughly equivalent to that of the European Union, but its position is slipping. In addition, if China continues the current rate of increase in publication numbers, its output will equal or surpass the United States within the next 5 years.

Panel's Summary Assessment. The current U.S. position is at the "Forefront," and in the future, it will be "Among World Leaders."

4.5 AREA-5: MATERIALS

How to design, make, use, and adapt materials, has been central to social advancement and economic growth since the dawn of history. Since the end of World War II, there has been an explosion in our understanding and application of the science and engineering related to materials. Many disciplines, e.g., chemistry, physics, materials science and engineering, chemical engineering, biology, and mechanical engineering, have contributed actively in the research for new and better materials and their more efficient production. For chemical engineering research the area of materials has been of increasing emphasis over the last 30 years.

For the purposes of this report, the Panel divided the area of materials into the following four subareas:

- polymers
- inorganic and ceramic materials
- composites
- nanostructured materials

Research in biomaterials and materials for cell and tissue engineering are discussed in Area-6 (see Sections 4.6.b and 4.6.c).

4.5.a Polymers

While most people equate polymers simply with plastics, these versatile and diverse materials are critical to the manufacture of an enormous range of products that include semiconductor chips, medical and pharmaceutical products, food packages, structural materials, materials for automobiles and airplanes, adhesives, paints, many other types of protective and functional coatings, and numerous consumer and household items. Quite simply, polymers are everywhere. Major U.S. corporations derive significant profits from the sale of polymers, formulated polymer systems, and downstream products which are enabled by polymers.

U.S. Position. The United States has had a historical leadership position in the field of polymers. Many of the most significant polymer development efforts in the United States were initiated before World War II, such as nylon and synthetic rubber. Polymer foam was also developed during World War II because of the need for flotation devices. In the 1950s, the United States expanded its leadership position, especially in high-performance polymeric systems. One of the most celebrated Nobel Prizes was awarded to Paul Flory of Stanford University in 1974 for his work in polymers. More recently, U.S. scientists Alan MacDiarmid and Alan Heeger received the

2000 Nobel Prize for work in conductive polymers. In the area of polymers for medicine, the United States has been in the forefront of development and commercialization for the past 50 years.

Polling of 20 world leaders in polymers resulted in the identification of 151 unique Virtual World Congress speakers, 68% of whom were U.S. based. On the basis of a total name count of 341, U.S.-based experts were more likely to receive multiple votes from the 20 experts, since the nonunique name count was 74% for U.S.-based Virtual World Congress speakers (see Table 3.1). Detailed analysis of the names of the speakers shows that the European Union and Japan make up most of the non-U.S. speakers.

Despite the strong interdisciplinary nature of this area (chemical engineering, materials science, and chemistry), there was a significant presence of chemical engineering speakers (about 20%). It is also worth noting that another 8% of these speakers were actually trained as chemical engineers but now work in materials science or chemistry departments.

The three top journals for polymer materials with significant impact are *Progress in Polymer Science*, *Macromolecules*, and *Polymer*. Of these, *Progress in Polymer Science* is a review journal, one of two major journals (the other being *Advances in Polymer Science*) that publishes exclusively invited reviews in the field. Analysis of each follows.

Tables 4.24 and 4.25 show the results for *Progress in Polymer Science*. Trends from this journal should be taken cautiously because of the journal's very low publication rate from chemical engineers (~5%) and small absolute number of publications (<15 in each 5-year period). From the 1990s through today, total U.S. contributions from all disciplines have doubled in number, while the fraction of the total has remained about the

TABLE 4.24 Papers Published in *Progress in Polymer Science*

	1990-1994		1995-1999		2000-2006	
		%		%		%
Total Number of Papers	133		149		238	
Total No. of U.S. Papers	21	16.00	36	24.00	47	20.00
Total No. of Chem. Eng. Papers	7	5.26	5	3.36	13	5.46
U.S., Chem. Eng.	1	14.29	2	40.00	4	30.77
EU, Chem. Eng.	2	28.57	0	0.00	4	30.77
Asia, Chem. Eng.	4	57.14	1	20.00	5	38.46
Canada, Chem. Eng.	0	0.00	2	40.00	2	15.38
S. America, Chem. Eng.	0	0.00	0	0.00	0	0.00
Internationalization (overlap)		0.00		0.00		15.38

TABLE 4.25 Distribution of the 30 Most-Cited Papers Published in
Progress in Polymer Science

	1990-1994	1995-1999	2000-2006
No. of U.S. Papers	4	3	3
No. of Chem. Eng. Papers	3	0	4
No. of U.S. Chem. Eng. Papers	0	0	0
(% share among chemical engineering papers)	(0%)	(0%)	(0%)

same. Worldwide chemical engineering contributions have also doubled in
number, with the fraction of the total remaining roughly constant. Finally,
U.S. chemical engineering contributions have doubled in fraction showing
a strong sign of increasing activity, but we are not yet seeing any significant
impact from this increased activity, because there were no U.S. chemical engineering contributions in the list of 30 most-cited papers (Table 4.25). It is
again noted here that these observations are for a journal that is publishing
invited reviews rather than original scientific discovery.

Tables 4.26 and 4.27 show results for the journal *Macromolecules*.
Data from this journal is more relevant than *Progress in Polymer Science*,
because it publishes original research results and the number of chemical
engineering contributions is larger (>500 chemical engineering contributions in each 5-year period) and statistically significant. As we can see from
Table 4.26, the fraction of U.S. papers has declined from 51% (1990-1994)
to 38% (2000-2006), indicating the increasing research output of polymer
science and engineering in other countries, especially Japan, Korea, and
China. The contributions from chemical engineers worldwide have in-

TABLE 4.26 Papers Published in *Macromolecules*

	1990-1994		1995-1999		2000-2006	
		%		%		%
Total Number of Papers	4,756		5,723		8,307	
Total No. of U.S. Papers	2,430	51.09	2,369	41.39	3,168	38.14
Total No. of Chem. Eng. Papers	537	11.29	772	13.49	1,365	16.43
U.S., Chem. Eng.	440	81.94	579	75.00	909	66.59
EU, Chem. Eng.	60	11.17	115	14.90	245	17.95
Asia, Chem. Eng.	72	13.41	130	16.84	354	25.93
Canada, Chem. Eng.	20	3.72	25	3.24	82	6.01
S. America, Chem. Eng.	1	0.19	1	0.13	6	0.44
Internationalization (overlap)		10.43		10.10		16.92

TABLE 4.27 Distribution of the 50 Most-Cited Papers in
Macromolecules

	1990-1994	1995-1999	2000-2006
No. of U.S. Papers	17	25	25
No. of Chem. Eng. Papers	8	2	16
No. of U.S. Chem. Eng. Papers	7	2	12
(% share among chemical engineering papers)	(87%)	(100%)	(75%)

creased substantially (>2X) and by almost 50% as a fraction of the total.
Thus, *Macromolecules* is of strong interest to chemical engineers, and this
interest is growing. It is worth noting, however, that the contributions of
chemical engineers in this journal are more in the areas of synthesis, physi-
cochemical analysis, kinetics, property estimation, dynamic behavior, and
molecular modeling and much less in polymer engineering and processing.
The U.S. chemical engineering contributions more than doubled, but the
relative amount decreased from 82% (of all chemical engineering contributions) in 1990-1994 to 67% in 2000-2006. Asian (including Japanese)
contributions increased 5-fold, and their relative amount doubled.

Relative to the impact of these publications, the United States leads
the most-cited list (50%), and U.S. chemical engineers dominate the list
of most cited among chemical engineering contributions (>75%) in each
period analyzed (Table 4.27).

Tables 4.28 and 4.29 show the results for the journal *Polymer*. Like
Macromolecules, this is a journal with many chemical engineering contributions. Although the number of U.S. publications has increased by about

TABLE 4.28 Papers Published in *Polymer*

	1990-1994	%	1995-1999	%	2000-2006	%
Total Number of Papers	3,010		3,893		6,654	
Total No. of U.S. Papers	788	26.17	727	18.67	1,423	21.38
Total No. of Chem. Eng. Papers	311	10.33	566	14.54	1,105	16.61
U.S., Chem. Eng.	185	59.49	208	36.75	395	35.75
EU, Chem. Eng.	24	7.72	75	13.25	123	11.13
Asia, Chem. Eng.	76	24.44	283	50.00	576	52.13
Canada, Chem. Eng.	23	7.40	24	4.24	78	7.06
S. America, Chem. Eng.	1	0.32	5	0.88	11	1.00
Internationalization (overlap)		−0.64		5.12		7.06

TABLE 4.29 Distribution of the 30 Most-Cited Papers in *Polymer*

	1990-1994	1995-1999	2000-2006
No. of U.S. Papers	18	13	15
No. of Chem. Eng. Papers	9	3	12
No. of U.S. Chem. Eng. Papers	7	2	9
(% share among chemical engineering papers)	(78%)	(66%)	(75%)

80%, its relative fraction of the total has slightly decreased. The chemical engineering contributions have tripled worldwide, raising the fraction of the total from 10% (1990-1994) to 17% (2000-2006).

U.S. chemical engineering contributions have more than doubled, but their relative fraction has decreased from 59% (1990-1994) to 36% (2000-2006). Asian chemical engineering contributions (including Japan's) have increased 7-fold and dominate the volume of contributions.

Regarding impact in this journal, U.S. contributions dominate the list of most cited, and U.S. chemical engineering contributions have a very strong showing in the list of most cited (about 30%) and dominate (75%) contributions from chemical engineers worldwide.

Relative Strengths and Weaknesses. The Virtual World Congress analysis shows that today the United States is a world leader in this field with ~70% of the top experts. In addition, U.S. leaders are more likely to be chosen by multiple organizers. This is clearly a strength for the United States at this point in time. Trends in publications, however, suggest that this position is at risk given the explosive growth in quantity and steady improvements in quality of the Asian research efforts.

Publication analysis shows that for the two premier journals in the field that publish original publications, more and more chemical engineers are publishing in the area of polymer materials. However, as a fraction of the total, U.S.-based authors from all disciplines as well as U.S.-based chemical engineering authors are losing significant share as publications increase from Asia.

On an impact basis using citation analysis, the picture is somewhat better. For both *Macromolecules* and *Polymer*, the fraction of the 50 most-cited papers that are coauthored by U.S. authors from all disciplines is strong. U.S. chemical engineers have an even higher fraction of the worldwide chemical engineering contributions. While this is a current strength, the overall trend of total article share loss is likely to change this picture in the future unless there is an increasing focus on the impact of future work. Quality and not quantity will have to be the approach if the United States is to retain leadership and influence.

Future Prospects. The polymer materials market is large. Through company consolidations and emergence of new applications the field continues to flourish. Advanced research on polymers is expected to increase over the next 20 years with the advent of combinatorial chemistry methods; fast-throughput techniques for rapid property characterization; new and improved technologies of polymerization; and advanced techniques of molecular structure and surface modification by functional group decoration, grafting, and other approaches. Such methods are expected to continue to drive development of new polymers into important areas such as electronics and health care and open new possibilities for polymers in applications such as affordable consumer and industrial products for emerging economies. We expect that major growth areas for polymer applications are separation media, barrier coatings, packaging, and electronic-photonic applications such as displays and resists. In addition, cost-competitive block copolymers, self-assembly and forced-assembly polymer technologies, polymers from biomass, nanostructured self-assembled polymers, polymers for portable power (fuel cells and batteries), holographic storage polymeric materials, advanced conducting and semiconducting polymers for electronic applications, advanced polymers with dynamic response, and self-healing polymers are some of the research challenges to be addressed by future research.

Although new functions are being continuously demanded of polymers, today there is little focus on new classes of polymers (as there was during the first half of the 20th century), and the emphasis is on more specialized polymers that are often simply "offspring" of current polymer platforms. Very important examples of this lie in the fabrication of microfabricated and micropatterned devices and semiconductor chips. The enabling lithographic manufacturing process is totally dependent on new polymers. This will drive more innovation and scientific discovery in academic, government, and industrial laboratories around the world. The increasing participation of chemical engineering contributions in the leading polymer journals is also an indication that this group believes in the importance of this area. Rising energy costs encourage research in biomass-based production of monomers and polymers.

The Panel's analysis clearly indicates the United States is currently in a leadership position at the "Forefront." However, emerging economies such as China's threaten to dilute the influence of the U.S. contribution, which would have a serious economic impact. As manufacturing jobs continue to migrate to Asia, the United States will be challenged to retain the more valuable jobs that make new scientific discoveries and ultimately turn them into new products. This can only be done by continuing to do high-impact work. This must be a top priority that is supported by adequate funding, which will also attract the best talent.

The Panel believes that in the future the United States will be "Maintaining" its current position at the "Forefront," but it assumes that U.S. government, academic, and business leadership understand the importance of this field and will ensure we educate and train the appropriate talent in this area. The Panel is cautiously optimistic about this prospect, and that is why we believe the United States will be "Maintaining" its current position. An example of where this is happening is in the new field of nanostructured materials (see below). Polymer materials are a subset of this subarea (and vice versa), and the strong U.S. funding of nanostructured materials research strengthens the research base of polymer materials as well. Our note of caution is due to the fact that there are many "headwinds" in the face of this optimism. In addition to the issues already noted above, a recent report on materials science and engineering has underlined concerns about funding levels in the United States versus other countries. Creative solutions should be aggressively explored such as the highly successful SEMATECH government- and industry-funded precompetitive consortium (*http://www.sematech.org/corporate/history.htm*). Such a consortium could be used, for example, to fund precompetitive work on alternate feed stocks for polymers and monomers.

Panel's Summary Assessment. Currently, the U.S. position is at the "Forefront," and in the future, the United States will be "Maintaining" this relative position.

4.5.b Inorganic and Ceramic Materials

Inorganic materials cover an extensive range of applications. Significant emphasis in recent years has focused on nanotechnology, semiconductors, electronic materials, phosphors, magnetic materials, inorganic materials for catalytic and environmental applications, inorganic-organic hybrids and, to a lesser but still significant extent, biomediated inorganic synthesis of materials. Experimental synthesis, characterization and properties, modeling of materials formation processes, and development of a broad range of applications are the questions attracting the majority of current research efforts.

U.S. Position. The U.S. position in inorganic chemistry and materials research remains on par with the rest of the world, but is not dominant. The two most highly cited articles authored by chemical engineers dealt with inorganic materials. A survey of the key journals in this area (*Advanced Materials, Inorganic Chemistry, Chemistry of Materials, Materials Research Bulletins, Inorganic Materials*) reveals that the U.S. contribution from 1997-2005 has remained relatively constant at 30%-35%. Of the unique speakers proposed by the Virtual World Congress, 63% were from the United States. Analysis of U.S. patents in ceramics is consistent with the other metrics indicating 51%, 48%, and 49% of U.S. patents were assigned to U.S. companies in 1995, 2000, and 2004, respectively. The United States remains a strong contributor to many areas of inorganic materials. However, in the area of advanced ceramics, the United States is losing ground to Japan, Korea, and Germany. There was also a significant increase of activity in China, with publications from the mainland growing from 2% in 1997 to 13% in 2005.

Relative Strengths and Weaknesses. Key strengths of the U.S. research can be found in solid-state electronic materials, catalysts and supports, ceramic composites, and nanomaterials. Applications of these inorganic materials to aerospace, defense, armor, telecommunication, and data storage and transmission are among the areas impacted by U.S. strength in the area of inorganic materials. The United States has lost some leadership in the area of traditional solid-state synthesis, ceramic processing, and more traditional coordination chemistry.

The United States does not have any large dedicated institutes such as Japan's *National Institute for Research in Inorganic Materials*. U.S. research tends to be concentrated around applications. Infrastructure implications for inorganic materials include the need for many more energy-efficient and precisely temperature-controlled furnaces for synthesis, and a recommitment to the development of analytical characterization instruments for inorganic materials, such as significant reductions in beam- or spot-size to chemically characterize nanoscale domains and grain boundaries. Perhaps of greatest impact would be the development of high-throughput capabilities, which could synthesize an array of materials at temperatures of up to 1500°C and then exhaustively analyze them.

Korea and China are making significant advances in the areas of inorganic materials and metallurgy. Unlike the past, many foreign-born experienced graduates are now returning to their native countries, and research facilities are improving to rival those in the United States. In addition, countries in the European Union enjoy significantly greater opportunities for longer-range funding than seems to be experienced in the United States today.

In spite of the need for continued growth in inorganic materials, given their temperature stability, versatile chemical and physical properties, and independence from hydrocarbon as a feedstock, the disappearance of much of the U.S. steel and ceramics industries has led many students to pursue other areas of study. In an attempt to attract future students, many traditional ceramics and metallurgy departments morphed into materials science and engineering departments with only occasional pockets of strength in ceramics and metallurgy.

Future Prospects. The United States will continue in the near term to be a key contributor to the area of inorganic materials. The United States and Japan share leadership in ceramics used for their thermal, electric, and mechanical characteristics. However, the Japanese manufacturing advantage (which has an effect on engineering and research), reduced U.S. funding of basic engineering research, and a perception that other areas such as biotechnology offer more attractive opportunities, the leadership in ceramic materials that the United States has enjoyed is likely to continue to decline. An exception to this trend is in the area of nanocrystalline and nanoporous materials, whereby increasing efforts by chemical engineers have led to significant advances in novel catalysts, ceramic membranes, fuel cells, and optical/electronic/magnetic materials.

Panel's Summary Assessment. The current U.S. position is "Among World Leaders," and although in the future this position is expected to weaken, the United States will remain "Among World Leaders."

4.5.c Composites

Composites are heterogeneous materials generally consisting of a matrix and fillers. The most common example is "fiberglass" in which glass fibers, used for structural reinforcing, are held together by a thermosetting resin (e.g., epoxy). As the strength-to-weight ratio becomes increasingly critical, the use of carbon fibers is growing. High-performance applications, such as windmills, jet engines, and plane fuselages (e.g., Boeing's 787), can often achieve their design specifications only through the extensive use of composites. A small portion of this area concentrates on inorganic composites where the matrix and/or reinforcing filler is a metal or ceramic for high-temperature application. Composites for the automobile industry offer the promise of significant reductions in total weight, and this area of applications is one that is expected to flourish significantly in the future. Composites for building materials is another broad area of applications.

U.S. Position. The United States does not dominate, but continues to be a strong contributor in both the research and application of composite materials. A reduction in military and space research funding has affected R&D activities in advanced composites. U.S. industry has maintained some activity but of reduced intensity. Recyclability issues associated with thermoset resins, as well as complicated and costly manufacturing processes, have limited the growth of consumer applications. U.S.-based composite fabrication centers were leaders during 1980-1990, but have declined as funding has been reduced. The European Union has maintained investment and continues in a leadership role. European Union legislation and priorities have driven the need for lighter weight and greener products. There has been a resurgence of some U.S. activity, mostly linked to nanocomposites and bioscience with the possibilities for biocomposites.

Although 86% of the experts polled were from the United States, only 63% of the proposed participants in the Virtual World Congress were U.S. based, further illustrating the relative weakness of the U.S. research enterprise in this subarea.

A survey of the key journals in this area, *Polymer Composites, Composites Science & Technology Composite Structures, Advanced Materials,* and *Chemical Materials,* reveals that U.S. contributions from 1997 to 2005 have remained relatively constant at 31%-39%. The European Union was more dominant, producing 41%-49% of the publications. The largest change was China, which went from 2% of the publications in 1997 to 14% in 2005. Of the uniquely named speakers proposed by the Virtual World Congress, 70% were from the U.S. This validates that the work in the United States is of high quality. Analysis of patents issued by the U.S. Patent Office on composites also shows a consistent and high level of contribution from the United States: in 1995, 60%, in 2000, 53%, and in 2004, 54% of the patents issued in composites.

Relative Strengths and Weaknesses. Centers focusing on composite manufacturing processes are difficult to maintain and are declining in number. This is a serious weakness, given the strong and continuous support of such centers in other parts of the world. Nonetheless, there is significant investment in composite research at universities by the U.S. Department of Defense, and composite technology is finding growing use in both commercial and military aircraft.

As with most areas, the talent follows the funding. Significant technical depth remains in the United States and could be reapplied to address issues in composites if sufficient funding and interest were to develop. However, students and faculty are currently pursuing research interests in other areas to the detriment of the composite materials.

Future Prospects. Some of the most notable advances during the past 10 years are the following: expansion in carbon fiber usage; ambient temperature curing of composites through electron beams; ceramic matrix composites; advanced dielectric composites; and electrophoretic preparation of thin films. However, as the international benchmarking of U.S. materials science and engineering research[1] has observed, ". . .basic research into

---

[1] "International Benchmarking of US Materials Science and Engineering Research," Appendix B in *Experiments in International Benchmarking of US Research Fields,* National Research Council, National Academy Press, Washington, D.C., 2000.

composites at US universities is coming to a standstill as a result of the Department of Defense decision to strictly curtail university research funding in metal, polymer, and ceramic matrix composites. If this situation long persists, the US could forfeit its leadership role in composites." Without a cost breakthrough, composites may not become a big research and new business platform in the United States. Once a cost breakthrough occurs, a drive may become apparent, and big issues in fuel economy and emissions regulation could become the driver for composites. Indeed, while academic research is at low level, new developments in industry are spurring a series of applications, especially in transportation (e.g., automobiles, airplanes) and construction. Furthermore, the emerging field of nanocomposites may provide additional impetus for new research and markets in the field.

Panel's Summary Assessment. Currently, the United States is "Among World Leaders," and in the future, the United States will be "Maintaining" its relative position.

4.5.d Nanostructured Materials

A nanostructured material is generally considered to be any material that has a feature of interest in at least one dimension that is 1 to 100 nanometers in size, or "nanoscale." Nanoparticles, quantum dots, nanocapsules, nanocrystalline materials (e.g., metals and ceramics), nanocomposites with structures modulated in some way at the nanoscale, and nanoporous solids are the most common subjects in this area. Potential end uses are extremely broad and include electronics, transportation, energy, consumer products, catalysis, and medicine. This field is quite young in comparison with other areas reviewed in this report. Further, although nanostructures such as 65- and 90-nanometer transistor gates in microprocessor chips are in commercial production, discovery of captivating novel materials such as carbon nanotubes and cadmium selenide quantum dots has yet to achieve significant economic impact. Nanopowders of zinc oxide and silver are, however, finding their way into products.

The promise of this field has caused great interest among chemists, physicists, material scientists, electrical engineers, and chemical engineers. As pointed out in earlier sections of this chapter, chemical engineers have slowed their research activities and publishing in traditional areas such as separations, transport processes, and thermodynamics and increased it in other areas such as nanostructured materials.

U.S. Position. This young field has always had an international flavor, with the discovery of fullerenes (1981 by Kroto in the United Kingdom, and Curl and Smalley in the United States) and carbon nanotubes (1991 by Iijima in

Japan) doing much to stimulate activity. Another major milestone was when U.S. researchers at IBM used the scanning tunneling microscope in 1989 to write the letters "IBM" with xenon atoms. These and other discoveries helped create the field of nanostructured materials, which has had heavy participation by U.S. researchers throughout its evolution. The establishment of the U.S. National Nanotechnology Initiative (NNI) in 2001 has fueled much activity with a total funding of over $6.5 billion through 2007 on nanotechnology from this funding source alone.

Polling of 13 world leaders in nanostructured materials resulted in the identification of 123 unique Virtual World Congress speakers, 65% of whom were U.S. based. When multiple nominations of the same individual were allowed, 74% of the 208 nonunique nominations were U.S. based (see Table 3.1). Analysis of the names of the speakers shows that the European Union (especially Germany) and Asia make up most of the non-U.S. names.

Virtual World Congress polls show the United States to be a strong leader today (75%-80%), followed by Europe, especially Germany. Despite the dominance of materials scientists and chemists in this field, chemical engineers have a significant presence.

The top journals for nanostructured materials are *Nano Letters*, *Advanced Materials*, *Chemistry of Materials*, and *Advanced Functional Materials*. Tables 4.30 and 4.31 show the results for *Nano Letters* and *Advanced Material*, two journals with similar impact factors. The abrupt increase in the number of papers during 2000-2006 is due to the initiation of *Nano Letters* in 2001, which had 1,973 papers during this period. During 2000-2006, the sudden rise in the fraction of U.S. contributions is

TABLE 4.30 Papers Published in *Nano Letters* and *Advanced Materials*

	1990-1994		1995-1999		2000-2006	
		%		%		%
Total Number of Papers	720		1,284		4,948	
Total No. of U.S. Papers	152	21.11	334	26.01	2,285	46.18
Total No. of Chem. Eng. Papers	8	1.11	39	3.04	374	7.56
U.S., Chem. Eng.	5	62.50	33	84.62	271	72.46
EU, Chem. Eng.	2	25.00	4	10.26	30	8.02
Asia, Chem. Eng.	2	25.00	4	10.26	124	33.16
Canada, Chem. Eng.	0	0.00	1	2.56	5	1.34
S. America, Chem. Eng.	0	0.00	0	0.00	8	2.14
Internationalization (overlap)		12.50		7.69		17.11

TABLE 4.31 Distribution of the 50 Most-Cited Papers Published in *Nano Letters* and *Advanced Materials*

	1990-1994	1995-1999	2000-2006
No. of U.S. Papers	14	21	29
No. of Chem. Eng. Papers	0	2	4
No. of U.S. Chem. Eng. Papers	0	1	2
(% share among chemical engineering papers)	(0%)	(50%)	(50%)

primarily due to *Nano Letters*, in which U.S. papers have accounted for roughly 66% of the total, while in *Advanced Materials* the U.S. fraction has been about 33%.

It is also clear that there is a very significant increase in the number (and fraction) of chemical engineering contributions worldwide. This is clearly a field that has seen growing interest from chemical engineers.

For U.S. chemical engineering contributions, we see a very strong increase over the past 10 years, both in absolute and relative numbers. The U.S. fraction of total articles appears to have peaked, however, as Asian chemical engineering authors have settled at around one-third of the chemical engineering papers in the latest period. There has been noticeable improvement in the fraction of most-cited papers by U.S., chemical engineering contributors over the period of study.

Tables 4.32 and 4.33 show the results for *Chemistry of Materials* and *Advanced Functional Materials*, two journals with similar impact factors and highly relevant to this field. The abrupt increase in the number of

TABLE 4.32 Papers Published in *Chemistry of Materials* and *Advanced Functional Materials*

	1990-1994	%	1995-1999	%	2000-2006	%
Total Number of Papers	1,329		2,335		5,804	
Total No. of U.S. Papers	934	70.27	1,232	52.76	1,957	33.72
Total No. of Chem. Eng. Papers	115	8.65	206	8.82	469	8.08
U.S., Chem. Eng.	111	96.52	164	79.61	274	58.42
EU, Chem. Eng.	0	0.00	11	5.34	53	11.30
Asia, Chem. Eng.	6	5.22	35	16.99	192	40.94
Canada, Chem. Eng.	0	0.00	9	4.37	5	1.07
S. America, Chem. Eng.	7	6.09	5	2.43	8	1.71
Internationalization (overlap)		7.83		8.74		13.43

TABLE 4.33 Distribution of the 50 Most-Cited Papers Published in *Chemistry of Materials* and *Advanced Functional Materials*

	1990-1994	1995-1999	2000-2006
No. of U.S. Papers	40	40	30
No. of Chem. Eng. Papers	5	5	0
No. of U.S. Chem. Eng. Papers	5	4	0
(% share among chemical engineering papers)	(100%)	(80%)	(0%)

papers during 2000-2006 is due to the initiation of *Advanced Functional Materials* in 2001, which had 830 papers during this period. While the number of U.S. papers has doubled from 1990-1994 to 2000-2006, the U.S. fraction of the total number of papers has been decreasing continuously due to the significant increase in the number of papers from Asia and the European Union.

Chemical engineering contributions in *Chemistry of Materials* and *Advanced Functional Materials*, worldwide, represent a respectable 8%-9%, much like the participation rate in 2000-2006 in *Nano Letters* and *Advanced Materials*. U.S. chemical engineering contributions have more than doubled in number, but their fraction of the total chemical engineering contributions has decreased due to increasing competition from Asian countries. The number of U.S. publications in the top 50 most-cited has decreased in the past 5 years, as has the number of highly cited U.S. chemical engineering publications in these two journals.

Relative Strengths and Weaknesses. The Virtual World Congress analysis shows that today the United States is a world leader in this field with over 70% of the top experts. In addition, U.S. leaders are more likely to be chosen by multiple organizers, indicating that they are particularly widely recognized. This is clearly a U.S. strength. Trends in publications, however, suggest that this position may be at risk.

Publication analysis shows that for the top two journals, *Nano Letters* and *Advanced Materials*, there is a greatly increasing interest of chemical engineering researchers in this field. U.S. contributors make up the largest fraction of this chemical engineering interest, but over the period of 2000-2006 there has been a substantial growth in Asian and European Union chemical engineering contributions as well. The second pair of journals, *Chemistry of Materials* and *Advanced Functional Materials*, shows a striking loss in article share for all U.S. authors and for U.S. chemical engineering authors, despite the increasing number of publications.

Thus, the picture today is one of strength with concern about the future, as further erosion of publications shares, i.e., higher relative growth rates in research activities around the world, might lead to a loss in critical U.S. impact in this field.

Future Prospects. Today's most advanced semiconductor chips are built with nanoscale transistors. Their fabrication also relies on nanoscale powders that are formulated into slurries used to planarize individual circuit layers during manufacture. New sunscreens use nanoscale zinc oxide to give better protection, and silver nanoparticles are being incorporated into household appliances for germ control. Such examples are only the start of a long and prosperous road for this young area that will see significant scientific discovery and resulting development of many important new products.

The Panel's analysis shows that while the United States leads the world in this area, it will be challenged to retain its position. Asia is investing heavily and can be expected to take a significant position in the long term. However, many of the applications for nanostructured materials are being conceived in the United States and are used by the U.S. semiconductor industry, which is still the most advanced in the world. Continued high investment by the United States is critical to ensure future success in this important emerging area of science and technology.

In the future, the Panel expects that the United States will be "Gaining or Extending" its current position at the "Forefront," primarily due to the fact that there is a high level of investment in the United States in this field. Investment is coming in the form of government-sponsored centers and research grants, direct academic investments, and commercial R&D by large corporations and small venture investors in and outside the chemical industry.

Panel's Summary Assessment. The current U.S. position is at the "Forefront," and in the future, the United States will be "Gaining or Extending" its relative position.

4.6 AREA-6: BIOMEDICAL PRODUCTS AND BIOMATERIALS

Chemical engineering research in health care-related matters has at least a 40-year-long history. During this period we have seen an increased collaboration between chemical engineers and medical researchers in addressing significant issues and coming up with innovative products such as dialysis devices, drug targeting and delivery systems, biomaterials for catheterization, wound healing and protection, surgical instruments, cardiovascular ailments, lenses, orthopedic applications, and others.

For the purposes of this benchmarking study, the Panel divided this area into the following three subareas:

- drug targeting and delivery systems
- biomaterials
- materials for cell and tissue engineering

4.6.a Drug Targeting and Delivery Systems

The subarea of drug delivery has attained a prominent position in chemical and biological research over the last 40 years. From its relatively simple infancy as a subarea of pharmaceutical sciences and as a research subject addressing predominantly formulation aspects for small molecular weight drugs, it has matured into a field that addresses the design and deployment of advanced systems for the delivery of small molecules, peptides, and proteins. Corollary research issues include detailed analysis of transport processes in carriers and tissues; carrier/tissue and carrier/cell interactions; advanced methods of analysis of cellular behavior; drug and protein absorption (transport) mechanisms; and modeling, pharmacokinetics, and pharmacodynamics. Chemical engineering educational preparation and technical skills are ideally suited to address these research issues.

U.S. Position. Eleven experts, 8 of whom were from the United States, proposed 94 participants, 64% of whom were U.S. based. Chemical Engineers comprised 38% of the participants, with the rest being pharmaceutical scientists (22%), chemists (18%), and others (22%). This is an impressive number of U.S. chemical engineers, considering that the field of drug delivery and controlled release started as a subarea of pharmaceutical sciences, and a clear recognition of the contributions of chemical engineers in the field.

Drug delivery scientists disseminate their research in original publications and review articles. This is a very competitive interdisciplinary field, where early protection in the form of disclosures and patents is desired and in fact promoted, even in the academic sector. Drug delivery scientists publish in many of the high-profile journals, such as *Science, Nature, Proceedings of the National Academy of Sciences, Chemistry of Materials, Biomacromolecules,* and *Nature Drug Discovery.*

The large majority of drug delivery research contributions are published in several journals of the field. The leader among them, *Journal of Controlled Release,* has seen significant increases in the number of publications from 597 in the 1990-1994 period to 2,022 in the 2000-2006 period, a near 3-fold increase (Table 4.34). While the number of U.S. papers has increased nearly 2.5 times, the relative ratio has decreased from 43% (1990-1994) to 30% (2000-2006). The contributions from chemical engineers have increased 4-fold, but the relative fraction has remained about the same, at 7%-9%. Over the last 8 years U.S. publications have doubled

TABLE 4.34 Distribution of Publications in the *Journal of Controlled Release*

	1990-1994	%	1995-1999	%	2000-2006	%
Total Number of Papers	597		936		2,022	
Total No. of U.S. Papers	257	43.05	346	36.97	606	39.97
Total No. of Chem. Eng. Papers	44	7.37	85	9.08	161	7.96
U.S., Chem. Eng.	27	61.36	54	63.53	98	60.87
EU, Chem. Eng.	3	6.82	16	18.82	24	14.91
Asia, Chem. Eng.	8	18.18	20	23.53	42	26.09

in number (from 172 in 1997 to 347 in 2005), but the corresponding fraction has decreased from 38% to 27%. The data clearly show that the rest of the world is publishing more in this field with China being a major new contributor (0 publications in 1997, 2 in 2000, and 29 in 2005). The number of publications directly associated with chemical engineers has been around 8% to 10%.

In terms of quality and impact, Table 4.35 summarizes the main results: the United States, the European Union (including Switzerland for this categorization), and Asia (primarily Japan and Korea) approximately share the fractions of most-cited papers, e.g., 10/10/10 (1995-1999) and 13/9/8 (2000-2006).

Three more traditional pharmaceutical journals that publish not only drug delivery papers but also papers in pharmaceutics, pharmacokinetics, in vitro/in vivo correlations, etc., are *Pharmaceutical Research*, *European Journal of Pharmaceutics and Biopharmaceutics*, *European Journal of Pharmaceutical Sciences*. The U.S. contributions in these journals in the 2000-2006 period were 50%, 14%, and 12%, respectively. Chemical

TABLE 4.35 Distribution of the 30 Most-Cited Papers in the *Journal of Controlled Release*

	1990-1994	1995-1999	2000-2006
No. of U.S. Papers	10	10	13
No. of Chem. Eng. Papers	0	2	3
No. of U.S. Chem. Eng. Papers	0	2	2
(% share among chemical engineering papers)		(100%)	(66%)
EU Chem Eng. Papers	10	9	9
Asian Chem. Eng. Papers	11	11	8

engineers contributed about 4% of the total, indicating their preference for journals with more chemical or technical rather than pharmaceutical orientation.

Finally, the leading review and assessment journal in the field, *Advanced Drug Delivery Reviews*, with an impact factor of 7.189, published 113 reviews in 2005, of which 42 (37%) were from the United States. The editorial practice of this journal (invited thematic review issues) does not allow for a totally independent analysis of the contributions, but scientists affiliated with chemical engineering have contributed 5% of the articles.

Relative Strengths and Weaknesses. The United States has had a historical leadership position in the field of drug delivery, and chemical engineers were a pivotal force behind the explosive developments of the 1970s and 1980s. They provided the scientific and methodological scope for principles-based rational frameworks in the design and optimization of what is now known as *system-responsive medical devices.*

This leadership is still in force, as reflected in the publications and citations record, and even more important, in the significant number of startups and the proliferation of very creative drug delivery systems in the market place by U.S. corporations. However, this is an area of significant interest and attention around the world. The European Union, Switzerland, and Asia have excellent research programs in drug delivery, as manifested by their strong presence in the list of the most-cited papers, and the levels of investment and research activity are growing strongly. The competition has been on for some time and will continue to be sharpened in the future.

Future Prospects. With the sales of advanced drug delivery systems in the United States approaching $20 billion annually, extensive research that focuses on improving and creating advanced drug delivery systems will continue. A significant portion of this market will continue to focus on the development and commercialization of "conventional" and generic drug delivery systems (tablets, capsules, micropowders), which are not as research-intensive. However, the development of advanced drug targeting and delivery systems will continue to be a real need and will provide an additional surge in associated research over the next 20 years. It is interesting that of the 2,698 original articles published in *Science* in 2005, 765 referred to "drug delivery" either directly or as a possible application of the published research.

The envisioned systems will provide a form of "intelligent response" as they do not simply release a drug at a specific rate, but release it to a specific site, often in pulses or in response to high concentrations of undesirable compounds. Additionally, because drug delivery can improve safety, efficacy, convenience, and patient compliance, improving delivery methods will become a major focus of pharmaceutical companies' research.

In recent years, microfabrication technologies have been applied in drug delivery, facilitating novel advanced drug delivery microsystems. These microfabricated drug delivery devices enable tailored drug delivery that is essential for the successful therapeutic activity of a drug. Although still in its infancy, this technology has demonstrated immense potential for surmounting barriers that are common to traditional drug delivery technologies.

Chemical engineers are uniquely qualified to address the drug targeting and delivery problems because of their education on chemical and biological processes and materials. As the needs in this subarea become more sophisticated, so will the research challenges leading to further expansion of interdisciplinary research opportunities.

Panel's Summary Assessment. The current U.S. position is at the "Forefront," and in the future, the United States will be "Maintaining" this relative position.

4.6.b Biomaterials

Biomaterials science and engineering started in the United States immediately after World War II, in response to the growing needs for materials compatible with the human body, e.g., medical devices such as artificial kidneys, contact lenses, and orthopedic applications. While most people equate biomaterials simply with artificial organs, and indeed a range of materials are critical to many aspects of reconstructive medicine, e.g. the manufacture of contact and intraocular lenses, artificial joints, assist devices, heart muscles, liver tissues, etc., the definition of this subarea extends to include biomedical drug delivery systems, such as insulin pumps, and other applications involving materials in the human body. Chemical engineers have played a pivotal role through their contributions in designing new biomaterials, composing improved evaluation methods of their biocompatibility, pursuing advanced understanding of material/tissue interactions, and catalyzing the use of biomaterials for a wide range of applications. Numerous U.S. corporations have established strong commercial leadership in the biomaterials field, a field with a global market in excess of $65 billion.

U.S. Position. The U.S. position in this subarea is very strong. Polling of 10 world leaders in biomaterials for the Virtual World Congress resulted in the identification of 77 unique speakers, 81% of whom were U.S. based. 52% were chemical engineers. Further analysis of the names of the speakers shows that the European Union (mostly France, Italy, and the Netherlands), Japan and Korea make up most of the non-U.S. speakers.

The two leading journals in the field, the *Journal of Biomedical Materials Research*, the official organ of the U.S. Society for Biomaterials, and *Biomaterials*, the official organ of the European and Japanese societies are comprised of 44% and 29% U.S. articles, respectively, and chemical engineers have contributed 10% and 12% of the papers. This is a healthy presence of chemical engineers in a field with many contributing sciences and engineering disciplines. The editors of *Biomaterials* from 1982 to 2002 were two U.S. chemical engineers. The *Journal of Biomaterials Science, Polymer Edition*, whose editor is a U.S. chemical engineer, contained 28% U.S. publications in 2005. The chemical engineering contributions were 19%.

A further analysis of the 100 most-cited chemical engineering publications in the period 2000 to 2006, revealed six publications of biomaterials content, with two publications in the top 10. A more detailed analysis of all 2000-2006 publications in the same archival source that listed "biomaterials" as a portion of its studies indicated that of the six most-cited scientists in the field, four are chemical engineers.

Relative Strengths and Weaknesses. The subarea of biomaterials is an interdisciplinary one, involving chemists, chemical engineers, materials scientists and engineers, biomedical engineers, biologists, and medical professionals (Japan and Korea). In the United States it has been populated and directed by many chemical engineers. From the early days, U.S. chemical engineers provided direction and leadership in basic and applied research (e.g., biomedical membranes functioning as separators in artificial kidneys in the mid 1960s), founded and managed the early corporate entities in this market, and defined the path that was followed in the subsequent 30 years of developments.

The establishment of the *Society for Biomaterials* in the United States was an important catalyst for the rapid advancement of the field. The Society had its first meeting in Clemson, SC, in April 1974, but was not incorporated as a Society until 1975. In the past 33 years, 18 academic and industrial chemical engineers have served in leadership positions of this organization.

A major impetus in this field was the establishment of federal funding in the United States by the National Institutes of Health in 1968. For the next 20 years, ample federal funding led to major research contributions in the fields of soft material replacement, biocompatibility, non-thrombogenic biomaterials, orthopedic biomaterials, and advanced composites. While countries such as France, Japan, Germany, Italy, and the Netherlands attained prominent positions in the world of biomaterials research by 1985, the United States became the leading country in the field and in related commercial ventures.

Biomaterials are essential for the future of the U.S. competitiveness in health care. As with drug delivery, the international competition has increased substantially. European Union and Asian investments in research and business development are significant and constitute a very visible threat to the U.S. preeminence in this subarea.

Future Prospects. Recent developments, inventions, and commercial successes required the use of advanced materials for biomedical applications. Indeed, biomaterials can be found in about 7,700 different kinds of medical devices; 2,500 separate diagnostic products; and 39,000 different pharmaceutical preparations. Just in the United States, the estimated annual sales of medical devices and diagnostic products in 2006 will be about $32 billion. Although biomaterials already contribute to an enormous improvement of health, there is a need for design and production of better materials with improved properties, with ability to have versatile functions, and with lower cost.

The development of biomaterials has been an evolving process. Many biomaterials in clinical use were not originally designed as such but were off-the-shelf materials that clinicians found useful in solving a problem. In the past few years, imaginative synthetic techniques have been used to impart desirable chemical, physical, and biological properties to biomaterials. Materials have either been synthesized directly, so that desirable chain segments or functional groups are built into the material, or indirectly, by chemical modification of existing structures to add desirable segments or functional groups. The advent of novel biohybrids will further fuel research activities in the synthesis, development, and commercialization of novel biomedical materials.

Our analysis indicates that the United States retains a leadership position in the field and that the younger generation of leading chemical engineering biomaterials scientists has both the reputation and recognition to direct the field for a substantial period in the future.

Panel's Summary Assessment. The current U.S. position is at the "Forefront," and in the future, the United States will be "Maintaining" its relative position.

4.6.c Materials for Cell and Tissue Engineering

Tissue engineering is a relatively recent field with about 20 years of research activities. Its objective is to develop the scientific understanding associated with the formation of cell tissues and convert this understanding into practical technologies, which would allow the eventual in vivo growth of human tissues for reconstructive and therapeutic purposes. Its starting point is usually associated with the first request for research proposals issued by the National Science Foundation in 1986. The early engineering was done in departments of chemical engineering and bioengineering. Chemical engineers have played a leading role in defining and advancing this field, as well as educating a generation of chemical engineers in this subarea. While the field has intellectually matured over the past 20 years, its scientific promise has not been translated into commensurable commercial success. Thus, although the scientific support by the National Institutes of Health continues, and although scientific symposia continue to be organized with great success, the industrial implementations are not yet evident.

U.S. Position. Eight experts identified 94 participants for the Virtual World Congress. The U.S. participation was very strong, with 78% of the nominations being for U.S. scientists and engineers (when duplication of nominations was allowed). This number was slightly reduced to 75% when duplication of nominations was disallowed. Of the 70 U.S. participants, it is interesting to note that only 16 were chemical engineers (23%), but an additional 30 participants (43%) have had chemical engineering education or were associated with chemical engineering units in the past, although they have since moved to other disciplinary units, such as biomedical engineering.

The leading journal in the field is *Tissue Engineering*, edited by a U.S. chemical engineer. It published 51% U.S. articles in 2005. About 18% of the published articles were by chemical engineers. Other journals publishing work by tissue engineers are covered in Section 4.6.b on Biomaterials.

Analysis of the 100 most-cited papers in chemical engineering from 2000 to 2006 indicates that there were seven papers from the field of tissue engineering.

Relative Strengths and Weaknesses. The international Tissue Engineering Society was formed in 1995. U.S. chemical engineers have been quite prominent in the leadership of this organization and have defined the scope of the field through their editorship of the leading journal, *Tissue Engineering*. While tissue engineering started in the United States, the strong medical component of this field has led to a rapid expansion to other countries and to medical schools. Engineers continue to be important participants in meetings and scientific organizations, but they do not have the prominence they once had. The same can be said about the position of U.S. chemical engineers in the field. Most experts expect that reconstructive medicine is one of the most promising areas for business development in the health-care field. Tissue engineering is the core of all such technologies. Chemical engineering has been critical in important developments, but its role seems to have been diminished. This is a threat and an opportunity for the future of chemical engineering at large, and U.S. chemical engineering in particular.

Future Prospects. In the past few years the subarea of cell and tissue engineering has adopted the more general scope of *regenerative medicine*, indicating an expansion to its repertory of materials and methods that will lead to tissue replacement. Currently, research activities utilize synthetic extracellular matrices to synthesize or regenerate tissues and organs. The materials that form the scaffold must be biocompatible, promote cell adhesion and growth, and biodegrade into nontoxic components. These have included poly(lactic acid), poly(glycolic acid), poly(DL-lactic-co-glycolic acid), and collagen-based matrices. They have been produced in a number of ways including freeze drying (collagen-based matrices for skin regeneration), fiber bonding (PGA, PLLA fibers for hepatocytes), foaming, salt leaching, and three-dimensional printing. Therefore, to create functional tissues, the key factors are extracellular matrices that anchor, orient, and deliver cells; bioactive factors to provide instructional and molecular cues; and cells that are capable of responding to their environment and capable of synthesizing the new tissue or organ of interest. However, the Panel has recognized that major commercial breakthroughs are needed in order to maintain the present levels of research interest and rekindle interest in additional independent funding.

Panel's Summary Assessment. The current U.S. position is at the "Forefront," and in the future, the United States will be "Maintaining" its relative position.

4.7 AREA-7: ENERGY

Energy provides the underpinning of an industrial society, with fossil energy dominating total energy consumption globally (~ 89%) and in the United States (~86%). Chemical engineers have had a long history of involvement in energy. The rapid growth of chemical engineering in the first half of the 20th century can be tied to the demand for technology and manpower by the petroleum and petrochemical industries. U.S. chemical engineers made major breakthroughs in the development of the processes for cracking, hydroprocessing, reforming, isomerization, coking, and distillation. As the industry matured and R&D departments downsized, the demand for chemical engineers declined, leading to a decline in the level of research activities with a corresponding decrease in publication rates. The recent increase in demand for petroleum, driven, in part, by the growing economies of China and India, and a decline in petroleum reserves, have led to a rapid escalation in oil and gas prices, and with them new opportunities for the chemical engineer: enhanced recovery of petroleum and gas; clean and efficient utilization of gas, oil, and coal; and development of major new industries for the production of liquid fuels from coal, shale, and tar sands.

In the longer term, as fossil energy resources become depleted, or their use possibly curtailed by penalties on the emission of CO_2, the need will be for non-fossil energy including solar, biomass, wind, and nuclear, as well as coal utilization with possible carbon capture and sequestration. For the purposes of this benchmarking exercise, the area of energy-related research is subdivided into the following three subareas:

- fossil energy extraction and processing
- fossil fuel utilization
- non-fossil energy

4.7.a Fossil Energy Extraction and Processing

The increase in the price of oil and gas has led to increased interest in enhanced oil recovery, liquefied natural gas (LNG), coal gasification, and coal liquefaction, all of which are technical areas that draw on the skills of chemical engineers. For example, CO_2 flooding (tertiary oil recovery) can extend the lifetime of mature fields, providing an extra 5%-15% recovery of the oil in the ground. With the prevailing high prices of petroleum tertiary oil recovery is again economical. Optimum strategies for recovering oil by the management of the fields requires sophisticated analysis of multiphase flow in porous media, which draws on traditional chemical engineering skills, skills that are also needed for the proposed sequestration of CO_2 in depleted gas and oil reservoirs and saline aquifers. The complementary increase in gas prices has led to revived interest in LNG imports to the United States, and the associated engineering of the gas purification, liquefaction, and revaporization plants. Also, the increase in gas and oil prices and the large U.S. coal reserves are motivating renewed interest in coal gasification and liquefaction.

U.S. Position. Virtual World Congress sessions were organized on reservoir engineering, LNG, coal gasification and coal liquefaction. The results are as follows:

- All of the speakers for reservoir engineering were from the United States, mostly petroleum engineers with a significant contribution (20%) from chemical engineers.
- For the LNG session all of the speakers were from the United States and from industry; where professional affiliation could be ascertained, they were mostly chemical engineers.
- The United States contributed 50% of the speakers for gasification, with most being chemical engineers.
- For liquefaction, 29% percent of the speakers were from the United

States. Japan had a stronger representation with 37%. The rest were from the European Union, Turkey, and South Africa. Chemists were dominant.

The United States was the major contributor to the U.S.-based journals related to petroleum and gas. For the *SPE Journal* the percent U.S. contribution declined from 71% in 1997 to 51% in 2005. U.S. contributions to *Oil and Gas* averaged 61%, when omitting an anomalously low 22% in 1997. The balance of the publications in both journals was from the European Union. By contrast the United States contributed an average of 23% of the articles in the *Journal of Canadian Petroleum Technology* with the majority (65%), not surprisingly, coming from Canada. Publications in *Fuel* cover the more fundamental literature related to coal science, gasification, and liquefaction. U.S. contributions averaged 12%, compared to 38% for the European Union. Contributions from China have exceeded those from the United States starting in 2003, and the gap between them is growing with time. The U.S. contribution to *Energy and Fuels*, one of the premier energy utilization journals, was 27% in 2005. In *Fuel Processing Technology*, another journal related to gasification and liquefaction, U.S. contributions have declined from 46% in 1997 to 25% in 2005. Again the major other contributors are the European Union and China, with an increasing contribution from India.

Relative Strengths and Weaknesses. The U.S. dominance of the oil and gas industries is not likely to diminish soon. However, there are problems of declining and aging manpower. The decline is both in chemical engineers and petroleum engineers, the latter being particularly in short supply reflected by a graduation rate in 2004 of only 20% of that in 1984. Industry hires and funding for research by both industry and government have been cyclical and these are reflected in the publication rates on gasification and liquefaction that have shown remarkable growth and decline in the 30-year period following the oil crises of the 1970s (see Figures 4.1 and 4.2).

For U.S. publications on gasification (Figure 4.1), one can note a peak in 1982 followed by a sharp downturn, with a leveling off around 2000 and moderate growth beyond 2002. More significant is the rapid rate of growth of gasification publications from China, which overtook U.S. publications in 2000. The U.S. contribution to the liquefaction literature (Figure 4.3) has shown a dramatic, near monotonic, decline since 1975.

The United States has been at the forefront of gasification, pilot units for Fischer Tropsch, and, through FutureGen, the world's first integrated sequestration and hydrogen production research power plant, a step towards fulfilling the vision for a hydrogen economy.

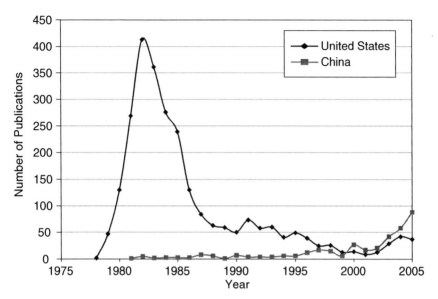

FIGURE 4.1 Publications related to gasification by United States and China for 1975-2005.
SOURCE: Data collected and presented by L. L. Baxter. Provided by L. L. Baxter, personal communication, 2006.

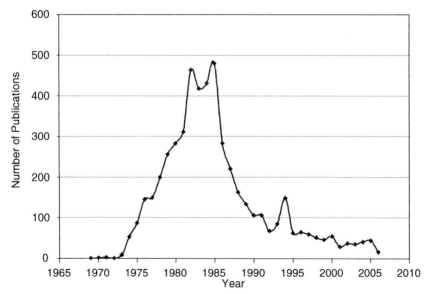

FIGURE 4.2 Trends in publications related to coal liquefaction for 1975 to 2006.
SOURCE: Data collected and presented by L. L. Baxter. Provided by L. L. Baxter, personal communication, 2006.

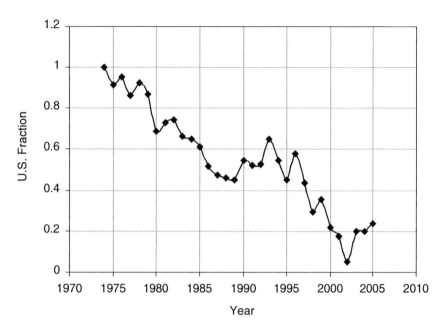

FIGURE 4.3 U.S. publications related to coal liquefaction as a fraction of total publications, 1974-2005.

Future Prospects. It can be reasonably anticipated that the oil price increases experienced during 2005-2006 will lead to a new resurgence in research on oil and gas, with improved exploration and recovery technologies for petroleum and natural gas, new technologies for coal gasification and liquefaction, and in situ recovery of oil from tar sands and shale. The difference in the activities this time will be greater international competition, particularly from Asia. The United States has a lead position in gasification technologies (GE-Texaco, Philips-Conoco, and Shell entrained flow gasifiers), with significant sales in China. Japan appears to be taking the lead in coal liquefaction, having operated a 150-ton/day pilot plant since 1992 and is aggressively pursuing opportunities in China. Competition from China will grow as China strives to transition from a manufacturing economy based on imported technologies to one based on domestic technologies. This is reflected by its investment in R&D and in their increasing rate of publication in the premier technical journals. Opportunities in which the United States has a major stake are the development of gasification, cleanup, and syngas utilization technologies that take advantage of developments in nanostructured materials. Carbon capture and sequestration could become a major area of activity, but the prospects of its adoption on a large scale are uncertain.

Panel's Summary Assessment. The current U.S. position is at the "Forefront," and in the future, the United States will be "Gaining or Extending" its relative position.

4.7.b Fossil Fuel Utilization

Fossil energy, used mainly in combustion devices, is the main provider of electricity, heating and cooling, motive power, and electricity in the United States. Nonenergy uses of fossil reserves for chemical production (e.g., coke from coal; petrochemicals, hydrogen, fertilizers from petroleum and natural gas) are relatively small. For combustion applications, the technical challenges are to achieve efficient, clean, and safe utilization for stationary sources. For motive sources additional challenges are ignition, flame stability, and flame blowout. The driving force for technology development continues to be tighter standards on emission of NO_x, SO_x, and particulate matter, using both combustion process modification and/or exhaust treatment. In propulsion systems challenges are in design and control of the new generation of engines, such as homogeneous charge compression ignition (HCCI) or scramjets, the latter requiring the ignition, stabilization, and completion of combustion at supersonic speeds. These processes are governed by turbulent reaction flows, an intersection of chemical kinetics and transport processes, made more difficult by short time constants and proximity to discontinuity in fuel conversion in time or space. Fossil energy utilization faces its greatest challenge with the rise of concern about global warming. The consequences of global warming are uncertain and more so are the consequences of the proposed mitigation strategies. If proposals to stabilize CO_2 emissions are implemented, they will require major developments of new fuel conversion technologies. Combustion and CO_2 mitigation technologies clearly draw on the core competencies of chemical engineers. Other disciplines actively involved are chemistry, materials science, and mechanical and aeronautical engineering.

U.S. Position. Sessions in the Virtual World Congress were organized on combustion science, technology, and policy; emissions from both automotive and stationary sources; clean and efficient power generation; micro and solid oxide fuel cells; ion transport membranes; carbon oxidation and gasification; and oxy-fuel combustion. Of the eight experts, five (62.5%) were from the United States. However, 55% of the nominated speakers were from the United States. The nominations to the Virtual World Congress from the three non-U.S. experts included only 41% of U.S.-based nominations; a measure of regional bias, as was seen in other subareas. Disciplinary affiliation was not always reported, particularly for industrial speakers. Chemical engineers represented 56% of the nominated participants. Interestingly, one expert inserted a policy dimension to the energy field—most appropriate at a time when many issues such as global climate and nuclear energy have clear social and political dimensions, and when the Secretary of Energy is a chemical engineer.

Analysis of the relevant journals on combustion and energy showed a decline in U.S. contributions to *Progress in Energy and Combustion Science*, a highly cited U.S.-based invited-review journal, from 100% in 1997 to 25% in 2005. Contributions from the United States to the major combustion journals showed no clear temporal trends and averaged 53% for the past two issues (previous issues were not abstracted by the American Chemical Society) of the *Proceedings of the Combustion Institute*, the premier publication for combustion science, 48% for *Combustion and Flame*, and 37% for *Combustion, Science and Technology*.

Relative Strengths and Weaknesses. As evidenced by representation in the Virtual World Congress and in publications on energy and combustion, the United States has maintained a strong position in the combustion field. Additional measures of the quality of the U.S. research in combustion is the receipt by U.S. researchers of 29 of the 59 gold medals awarded by the Combustion Institute from 1958 (date of first awards) to the present. However, the United States has not maintained as strong a position in technological contributions. In a growing number of areas leadership has moved to the European Union and Japan. In the area of automotive engines, Japanese and European Union firms have taken the lead in the development of the hybrid-electric gasoline engine and common rail direct injection engine for diesels. For stationary combustion Japan is taking the lead in the introduction of high-efficiency electricity generation plants, using ultra-supercritical boilers. China, the European Union, and Japan are providing strong competition for leadership in fluidized bed technology. U.S. weakness in translating strengths in fundamental combustion science to applications is partly due to the perception that combustion and energy are mature technologies and that there is little need for applied research. The exception is in gas turbines and the development of the next generation of propulsion systems, where the applied efforts have been sustained by the interest of National Aeronautics and Space Administration and the Department of Defense. Although the United States has maintained a position among world leaders in clean combustion technologies, it has fallen behind in technologies for increasing the efficiency of energy utilization for both stationary and mobile applications.

Future Prospects. One can anticipate major transformations in energy utilization technologies in the next decades with drivers being the needs for increased efficiency and decreased emissions. Action on global climate change and any penalties imposed on carbon emissions will have the greatest impact. The leading contenders for mitigating carbon emissions from coal-fired utility boilers, one of the major sources of carbon emission, are in the near term (by 2050) integrated gasification combined cycle (IGCC) with water gas-shift and oxyfuel combustion. The United States is taking the lead for the former and the European Union, Japan, and Australia the latter. Development of these technologies will require major inputs from the engineering community with chemical engineers being particularly in demand for the gasification route. Future developments, such as the increase in conversion efficiency of fossil fuels to electrical energy using solid-oxide and proton-exchange-membrane fuel cells, chemical looping, advanced cyclic CO_2 absorption or desorption schemes, and the next generation of oxyfuel plants with oxygen transport membranes, also provide great opportunities for the engineer, particularly the chemical engineer. The development of gas turbines and fuel cells running on hydrogen and chemical and fuel synthesis from syngas will be important components of proposed polygeneration plants. The need to accelerate the translation to markets of technical innovations will be facilitated by advances in predictive science—at a molecular level for chemical rate constants, at a component level for engines and furnaces, and at a system level. Validation and verification will be important to the acceptance of simulations, which in turn will depend upon the advances in diagnostics and instrumentations using lasers and high-energy beams.

Panel's Summary Assessment. The current U.S. position is "Among World Leaders," and in the future, the United States will be "Maintaining" this relative position.

4.7.c Non-Fossil Energy

No one source can replace the gap resulting from diminishing fossil fuel supplies. The contenders are nuclear, which will require inputs from chemical engineering for fuel reprocessing; biomass, used in combustors, gasifiers, and as a source of biodiesels; geothermal; photovoltaics; and wind. Sources such as wind and solar are intermittent and require energy storage media such as batteries. Another problem is that of developing a high-energy density transportable fuel to replace fossil fuel derived liquids; hydrogen is being seriously considered for this purpose in the United States. The areas best aligned with chemical engineering skills are developing improved biomass transformation products as alternatives to fossil energy; improving the energy density and life cycle of batteries; and developing hydrogen separation technologies. Contributions by chemical engineers to other areas are in support of complementary disciplines: geologists for geothermal; a range of disciplines for photovoltaics, including chemistry, physics, and materials science; and mechanical, aeronautical, and materials scientists for wind.

U.S. Position. The range of technologies is broad, and the selection of the following topics for the Virtual World Congress is far from complete: biomass direct utilization; biomass gasification; biofuels and biomass-derived green chemicals; geothermal energy; electrochemistry (batteries); adsorption-enhanced hydrogen production. U.S.-based speakers constituted 48% of the total nominations for the Virtual World Congress. However, the recommendations of the non-U.S. experts (two out of six) included only 25% U.S.-based participants. Chemical engineers constituted 40% of the speakers over all, with a higher percentage of 50% for the biomass-related areas. The session on hydrogen separation was dominated by chemical engineers, whereas geothermal was dominated by petroleum engineers and batteries by chemists and material scientists.

Analysis of a small number of specialized publications showed that the U.S. contribution to *Biomass and Bioenergy* was variable and averaged 21% of the total. It was exceeded only by European Union contributions. Contributions to *Solar Energy and Solar Materials* averaged 11% and *Wind Energy* 15%. Chemical engineers contributed more than their peer groups (chemistry, biomedical engineering, biology, and materials science) to *Biomass and Bioenergy*. The sessions for the Virtual World Congress were limited, and the samplings of journals was too small to draw any firm conclusions, but they support the perception that chemical engineers are aligned best with biomass utilization, both direct and after conversion to syngas, green chemicals, and biodiesel.

Relative Strengths and Weaknesses. The United States has strong programs in biomass utilization but is competing with the European Union, Canada, and countries in tropical zones that have high yields of bioenergy crops, such as Brazil, where ethanol from sugar cane supplies 40% of the fuel that would be needed to run the transportation fleet on gasoline alone. The increased gas prices have contributed to the growth of interest in gasification of biomass including black liquor, paralleling the increase in interest in coal gasification. The United States will continue to face strong competition from Canada and the European Union in the biomass area. The growth in utilization in the European Union is driven in part by the financial incentives provided by carbon penalties. The issue on how well the United States will be able to maintain a strong position is driven to a large extent by political considerations.

Future Prospects. Renewable energy is an essential component of both sustainability and a partial solution to global warming (partial because it is unlikely that biomass can replace more than about 20% of fossil fuel consumption). However, much of the renewable energy (hydropower, tidal power, solar energy, wind energy, and biomass) depends on solar insolation, which is diffuse. Biomass utilization in its various forms shows the potential for rapid growth with active participation of chemical engineers. Progress has been made in the direct utilization of biomass usually co-fired with coal, production of biodiesel by the transesterification of rapeseed oil and yellow grease, and the production of ethanol from sugar cane and corn. The challenge for the future is to effectively use all of the ingredients of biomass in a forest or from crops to produce a variety of green chemicals in addition to heat and power, in what has been named a biorefinery in analogy to a petroleum refinery. The lead efforts on biorefineries are in the United States, Canada, and the Nordic countries. Chemical engineers are expected to play a major role in the development of the biorefinery and materials for multijunction photovoltaic cells. One example is the development of improved catalysts for chemical and biochemical conversion of lignin-cellulose biomass to fuels. Chemical engineering will play an important, but lesser, role in the development of other sources of renewable energy. The largest challenge with abundant room for leadership is advocating and genuinely supporting a plan that will decrease the dependence on fossil fuels, probably including the use of nuclear, taking full account of the problems related to emissions and waste disposal.

Panel's Summary Assessment. The current U.S. position is "Among World Leaders," and in the future, the United States will be "Maintaining" this relative position.

4.8 AREA-8: ENVIRONMENTAL IMPACT AND MANAGEMENT

Environmental impact and management is an interdisciplinary field to which chemical engineers make critical contributions. In addition to the traditional areas of water and air pollution, new challenges now include concerns about global climate change and of pollution prevention or green engineering. The United States maintains a healthy leadership position in the environmental field, with a strong and growing program even though the percentage of the total contributions is decreasing due to higher growth rates in Europe and Asia. Over the period 1997-2005, U.S. authors contributed 53% to 65% of the articles to the leading U.S. environmental journals: *Environmental Science and Technology* and the *Journal of Air and Water Management*. The U.S. contribution to *Chemosphere*, a journal based in Europe, was only 13% showing the preference of authors to publish in their regional journals. The number of publications in *Environmental Science and Technology* from the United States and the European Union increased from 355 to 688 and from 143 to 508, respectively. China and India showed very high rates of growth, with the annual contributions by China increasing from 2 to 81 over the 5-year period. On average, chemical engineers contributed 5.8% of the papers in *Environmental Science and Technology*, the most cited of the environmental journals, exceeded only by chemists (10%) and biologists (7.4%). Chemical engineers, however, have taken the lead in selected areas, such as modeling the fate and transport of pollutants, aerosol science and technology, and controlling pollutants at their source. The four subareas covered in the Virtual World Congress are:

- air pollution
- water pollution
- aerosol science and engineering
- green engineering

4.8.a Air Pollution

Air pollution deals with sources of air pollutants, their transport and transformation in the atmosphere, and their impact on health, the natural environment, and materials. On a decreasing spatial scale air pollution is concerned with global climate (due to the depletion of stratospheric ozone and global warming); emissions of the criteria air pollutants (ozone, particulate matter, nitrogen dioxide, sulfur dioxide, and lead); and the 188 hazardous (toxic) air pollutants emitted during manufacture and use of industrial chemicals. Chemical engineers contribute to the management of air pollution problems by

- controlling the production of pollutants through process changes and development of technologies for the separation or destruction of the pollutants;
- modeling the fate and transport of the pollutants, particularly the formation of undesirable by-products such as ozone and organic particulate, and utilizing the models to guide control strategies; and
- supporting toxicologists in providing pharmacokinetic models for the distribution of chemicals in vivo, materials scientists on the effect of chemicals on building materials and products, ecologists in understanding and minimizing the impacts of chemicals (e.g., acid rain) on crops and ecosystems, and archeologists in restoring historic artifacts and buildings.

U.S. Position. Seven experts, six from the United States, organized sessions on scrubbers, catalytic processes, and pressure and temperature swing absorption for control of emissions; environmental monitoring and modeling of urban, regional, and global air pollution; and the formation and health impact of fine particles.

•   The Virtual World Congress speakers from air pollution control technologies were nearly exclusively chemical engineers, with the possible exception of speakers from industry whose educational background was unknown. The national affiliation of speakers was 46% United States, 26% European Union, and 18% Asia.

•   The Virtual World Congress speakers from environmental monitoring and the modeling of urban, regional, and global air pollution were drawn from a variety of disciplines, with major contributions from atmospheric chemistry, atmospheric science, and civil and environmental engineering. U.S. participants, however, were dominant, representing 74% of the total, with European Union speakers representing most of the balance.

•   In the area of the health effects of pollutants, chemical engineers provide a key supporting role to toxicologists in identifying the complex mixture of chemicals and aerosols that are characteristic of toxic air pollutants. U.S. speakers constituted 63% of the total, with the balance being primarily from the European Union. Chemical engineers constituted more than a third of the speakers with the balance drawn from a wide range of disciplines, including mechanical engineering, public health, chemistry, and physics.

The publications of interest vary widely: *AIChE Journal, I&EC Research,* and *Environmental Science and Technology* for control technologies; *Environmental Science and Technology, Atmospheric Environment, Journal of Air and Water Management, Atmospheric Chemistry and Physics, Journal of Geophysical Research* for transport and fate; *Health Effects Perspectives* for health effects. U.S. contributions to air pollution control is covered in the chemical engineering journals reviewed elsewhere, which show a strong and growing number of publications by U.S. researchers, but a declining percentage of the total because of greater growth rates in the European Union and Asia. Contributions to the specialized air pollution journals, *Journal of Atmospheric Sciences, Journal of Geophysical Research,* and those relating to the health impact of air pollution, generally show a greater than 50% contribution by U.S. authors, but no major trends with time. Journals such as *Atmospheric Chemistry and Physics,* based in Europe, showed a smaller U.S. contribution of about 25%.

Relative Strengths and Weaknesses. The number of publications in air pollution has been increasing, with the United States maintaining a strong leadership position. However, the aggregate statistics can be misleading. There are two technological drivers for air pollution studies. The first is the establishment of the causal relationships between anthropogenic emissions and adverse health that lead to the establishment of regulations on emissions. The second is the development of the technologies to bring industry into compliance with the regulations. U.S. chemical engineers contributed to the interdisciplinary studies that led to the understanding of the causal relationship between emissions of sulfur oxides and particles and increased morbidity and mortality; the contributions of SO_2 and NO_2 emissions to acid rain; and photochemical transformation of hydrocarbons and nitrogen oxides into photochemical smog and ozone. In response, the United States took the lead in the establishment of the standards with the promulgation of the Clean Air Act of 1970. Once standards were established, U.S. chemical engineers took the lead in the development of a series of technologies for SO_2 control, NO_x control, and simultaneous control of NO_x and hydrocarbons from automotive sources. They have also taken a lead in the development of the models used to set up state implementation plans for controlling the emissions of hydrocarbons and nitrogen oxides to meet ozone standards at the urban and regional levels. The success of the emission reduction program is reflected in the decrease over the 30-year period 1970 to 2002 in the aggregate emissions of the six principal pollutants by 48%, despite increases in population of 38%, energy consumption of 42%, vehicle miles traveled of 155%, and gross domestic product of 164%. The United States is among the world leaders in flue gas desulfurization, catalytic processes for pollution abatement, and mercury control technologies. However, as the European Union and Japan have adopted more stringent emission standards, they have taken the technological lead in selected technologies, e.g., selective catalytic reduction (SCR) of NO_x.

The United States developed SCR, but it was commercialized in Japan and Europe; the United States is now importing SCR technologies to meet stringent regional emission regulations prompted by failure to meet local ozone standards. The European Union is taking the lead in waste treatment technologies motivated by the Landfill Directive of the European Union. The United States is falling behind in the implementation and realization into commercial practice of new pollutant control ideas. An exception is the lead being taken by the United States in regulating the emissions of new pollutants, e.g., mercury with the promulgation of the Clean Air Mercury Rule (CAMR; March 2005). As a consequence, research and publications on mercury control technologies are rapidly increasing in volume with the United States in a strong lead position.

Future Prospects. With increased population densities and per capita consumptions, there will be continued tightening of emission limits of regulated pollutants in the United States. More stringent controls of SO_x and NO_x emissions will be required as part of the Clean Air Interstate Rules promulgated in March 2005. The very low NO_x emission requirements in ozone nonattainment areas such as Houston have provided constraints on industrial operations as well as an impetus for development of a new generation of low-NO_x burners. One can also anticipate regulation on carbon emissions that will require major research activities to develop and implement mitigation strategies. The challenges also provide opportunities for export of technologies and consulting services to the emerging economies that often set up standards modeled on those adopted by the United States.

The greatest challenge, however, will come from any adoption of regulations for carbon emissions. The problem is of such magnitude that it will require the adoption of multiple strategies, including conservation, renewable energy, increased nuclear, carbon capture and sequestration, and multiple disciplines. The impact of global warming on urban and regional air pollution will require the active involvement of air pollution modeling and chemical engineers.

Panel's Summary Assessment. The current U.S. position is "Among World Leaders," and in the future, the United States will be "Maintaining" this relative position.

4.8.b Water Pollution

The subarea of water pollution in the United States is covered mainly by the civil and environmental engineers. The areas in which chemical engineers provide critical leadership are

• development of water purification technologies for multiple purposes, notably drinking, irrigation, and for specialty industries such as microelectronics that have very stringent standards. Technologies for water purification draw on traditional chemical engineering process development and implementation. Chemical engineers also contribute to the biotreatment of wastes together with civil, environmental, and bioengineers.

• assessment of the problems associated with the release of toxic chemicals into the natural environment and prediction and/or mitigation of exposure by humans or ecosystems. The need for these assessments is both prospective, in premarket screening of chemicals, and retrospective, in dealing with the adverse consequences of chemicals released into the environment. The fate of chemicals in the natural environment involves the multi-media partitioning of chemicals, transport through porous media, and occasionally two-phase flows. The assessment of the impact of chemicals in the environment is multidisciplinary, involving civil, environmental, and bioengineers, geophysicists, chemists, and toxicologists.

U.S. Position. Seven experts, all from the United States, organized sessions in the areas related to water purification (water purification, water quality management and control, water separation and desalination) and the fate and transport of pollutants (chemodynamics, environmental chemistry, environmental fate of organic chemicals, control of hazardous substances).

• The speakers in the water purification area were predominantly (91%) from the United States. Chemical engineers represented 58% of academic speakers and environmental engineers represented an additional 33%.

• The majority (78%) of speakers in the area of fate and transport of pollutants were from the United States, with the balance from the European Union (15%) and Canada (6%). The speakers in the fate and transport area were mostly (56%) associated with an environmental department, often joint with civil engineering and/or geography; chemical engineers represented 20% of the total.

Contributions to technologies for water purification are expected to be distributed among journals dealing with separation processes, covered elsewhere in this report, with indications of a strong U.S. position. Many of the articles on the fate and transport of contaminants are published in *Environmental Science and Technology*, the *Journal of Air and Water Management*, and *Chemosphere*, discussed at the beginning of this section. Publications specializing in water treatment show that the publication rate was fairly constant over the period 1997 to 2005 for the *Journal of Contaminant Hydrology*, *Ground Water*, *Water Science and Technology*, and *Water Resources Research*, as were the U.S. contributions. The United States had the largest contribution to the *Journal of Contaminant Hydrology* (50%), *Ground Water* (70%), and *Water Resources Research* (64%). However, the United States contributed only 11% of the papers in *Water Science and Technology*, an international journal with offices in London. The European Union had the largest contribution to *Water Science and Technology* (52%) and the second largest to the other journals.

Relative Strengths and Weaknesses. Water pollution problems probably influence the chemical engineer most in terms of the regulations pertaining to the development and use of chemicals. U.S. researchers, many of them chemical engineers, have led the development of models for predicting the risks from chemical releases due to production, processing, usage, and disposal, using structure-activity relationships to extrapolate the risks to new chemicals. The models provide the scientific base for the implementation of the Toxic Substances Control Act of 1972. Similarly, chemical engineers were involved in the assessment of transport of chemicals in ground waters, essential to the characterization of hazards of waste disposal sites, responsive to the enactment of the Comprehensive Environmental Response, Compensation, and Liability Act (CERCLA or Superfund) of 1980, and active in the development of technologies for the remediation of these sites. The weakness, if any, is that students do not appreciate the applicability of a classical chemical engineering education to environmental problems and that, although the fate and transport of chemicals are one of the constraints on the development of new chemicals, chemical engineers will play a diminishing role in addressing the development of new and more effective tools to deal with these problems.

Future Prospects. Exciting opportunities exist for chemical engineers in both the development of water purification technologies and in reducing the risk from chemicals released into the environment. The United States is facing increasing water shortages in the West and Southwest. New water purification technologies have a role to play in the treatment of water for both U.S. municipalities and for undeveloped countries, for which water pollution problems constitute the major source of disease and water availability a major hurdle to development. New technologies in combinatorial testing and microsensors and the molecular understanding of toxicology open up opportunities for development of less hazardous chemicals and more effective ways to reduce the risk from the release of chemicals in the environment. Chemical engineers can continue to play an important role in combination with chemists, toxicologists, and environmental scientists in reducing the time and cost of bringing new chemicals to market and the risk once they are introduced.

Panel's Summary Assessment. The current U.S. position is "Among World Leaders," and in the future, the United States will be "Maintaining" its relative position.

4.8.c Aerosol Science and Engineering

The aerosol community is diverse with wide-ranging interests related to health (the assessment of the hazards of inhaling fine particles as well as the use of aerosols for inhalation therapy); environmental impact of aerosols on human exposure, visibility, and climatic change; and the synthesis of aerosols for use in a wide range of products including pigments, planarization agents for microelectronics, composite materials, and others. Chemical engineers have played a key role in the characterization of aerosols, modeling the formation and evolution of size and shape, and incorporating these models into general formulations that describe environmental impact at the local, urban, regional, and global level. Chemical engineers interface with researchers from a wide range of disciplines including physicists, chemists, toxicologists, atmospheric scientists, and all branches of engineering. Chemical engineers have taken the lead in bringing the diverse communities together in founding the American Association of Aerosol Research (AAAR).

U.S. Position. The experts for the Virtual World Congress focused on two areas with the following results:

• *The formation of atmospheric aerosols and environmental consequences.* In this area the United States contributed 70% percent of the speakers, the European Union contributed 25%. The affiliations of the speakers varied widely, with the largest numbers being chemical engineers, chemists, and environmental/aerosol scientists, each with about 30% of the total. *Aerosol synthesis of nanostructured materials.* U.S. speakers at this Virtual World Congress represented 52% of the total, with participants from the European Union and Japan comprising most of the remainder with contributions of 31% and 26%, respectively. Again the speakers represented many disciplines with chemical engineers contributing 27% of the total, chemists 19%, physicists 14%, and other engineering disciplines and materials science 31%.

The publication rate in the journals dedicated to aerosols—*Aerosol Science and Technology* and the *Journal of Aerosol Science*—show a moderate increase during the period 1997 to 2005. U.S. authors contributed 66% of the papers in *Aerosol Science and Technology* (associated with the American Association of Aerosol Research) with no clear trend over time.

The second largest contribution was from the European Union; the contribution from Asia was small. By contrast the contributions of U.S. authors in the *Journal of Aerosol Science* (associated with the European Aerosol Assembly) averaged 34%, and were exceeded by those of European Union authors for all years excepting 2005. The percentage of papers attributed to chemical engineers, although small (<20%), exceeded those by chemists, biomolecular engineers, biologists, and materials scientists. The *Journal of Colloid and Interface Science, Journal of Nanoparticle Research,* and *Powder Technology,* even though not dedicated to aerosol research, were also surveyed because of their inclusion of many papers on aerosols. The U.S. contributions to these journals were 37% for the *Journal of Nanoparticle Research* with no trend for the brief period (2003-2005) surveyed; 23% for *Powder Technology*; and 21% for the *Journal of Colloid and Interface Science*. For the latter two journals the number of articles contributed by U.S. authors increased over the period surveyed, but the U.S. percentage decreased because of greater growth rates from the European Union and China. The increase in the rate of publication by Chinese authors in *the Journal of Colloid and Interface Science* from 30 in 1997 to 153 in 2005 is notable (U.S. numbers for these dates are 159 and 181).

Relative Strengths and Weaknesses. The United States is a world leader in aerosol science and technology. Many—probably most—government agencies and major labs have aerosol programs including the Environmental Protection Agency, Department of Defense, Department of Energy, National Aeronautics and Space Administration, National Institute for Occupational Safety and Health, Nuclear Regulatory Commission, National Institute of Standards and Technology, and National Center for Atmospheric Research. Many deal with "bad" (that is, undesired unintentionally produced aerosols with undesirable effects on the environment and public health). However, there are few formal cooperative efforts among these organizations. Many industries have major aerosol-based commercial activities, for example Cabot, DuPont, Dow, Corning, in addition to many start-ups that manufacture nanoparticles by aerosol processes.

Significant contributions have been made to defining the major uncertainty in the role of the atmospheric aerosol in climate change, the development of aerosol reaction engineering as a major design methodology for companies such as Cabot, DuPont, Degussa, Corning, ATT/Lucent, and major breakthroughs in aerosol instrumentation largely driven by academic researchers now marketed commercially (e.g., online differential mobility analyzers for particle size distribution measurements, online single particle aerosol chemical analysis by mass spectrometry, and an aerosol aerodynamic lens TSI).

This subarea of aerosol science and technology shows a healthy growth with continuing challenges in the environmental field and new challenges from threats of global warming and bioterrorism. The growth of industrial applications for nanostructured materials has provided opportunities for the synthesis of novel materials and new manufacturing techniques. The U.S. programs are strong and growing, with the United States contributing a greater number of publications to the lead journals in the area. The highest honor in aerosol research, the Fuchs Memorial Award, jointly administered by German, Japanese, and U.S. institutions and given every 4 years, was awarded to U.S. chemical engineers in 1990 and 1998, was shared by a U.S. mechanical engineer in 1994, and was jointly awarded to a U.S. chemical engineer and a U.S. mechanical engineer in 2006. Other awardees for 1994 and 2002 were from Austria, Japan, and Russia. The United States is facing increasing competition from the European Union and Japan in all aspects of aerosol research and technology, and it is anticipated that China will soon become a serious competitor, judging from the growth in the number of Chinese publications in the *Journal of Colloid and Interface Science*. The area of aerosol science and technology is highly interdisciplinary. Chemical engineers have traditionally produced the leaders in the field. It is hoped that this tradition can be maintained by highlighting the opportunities in the field to future generations of students and young faculty.

Future Prospects. The future of aerosol science and technology is bright, with opportunities arising in both the environmental and materials synthesis areas. Currently regulations on the health effects of fine particles are based on correlations from epidemiological studies between the mortality and morbidity and the mass concentration of particles smaller than 2.5 microns in diameter (called fine particles). Regulations have been promulgated that control the ambient concentration of particles under 2.5 microns in size, in the absence of evidence of the composition and the actual sizes of particles responsible for the observed health effects. The United States has taken the lead in establishing the importance of fine particles on health and is taking the lead in the characterization, both theoretically and experimentally, of smaller particles, including nanoparticles (1 to 100 nanometers) that many believe to be of primary concern. Similar challenges are present in characterizing the role of particles on global climate. Nonabsorbing particles, primarily sulfates and nitrates formed in the atmosphere from the emissions of nitrogen and sulfur oxides, have a negative radiative forcing tending to a cooling of the surface temperatures whereas carbonaceous particles (mainly soot) are responsible for a positive radiative forcing. The magnitude of the forcing functions are dependent on the size of the particles, the details of their composition (for mixtures of soot and condensate, whether homogeneous, coated, or mixed), and their distribution with height and altitude. These are challenges that draw on many disciplines, with chemical engineers playing a major role. The skills needed to address the formation and characterization of the complex environmental particles are the same as those required to synthesize nanoparticles of given size, shape, and composition, and many of the chemical engineers studying environmental aerosols also contribute to their synthesis. As the area of aerosol, science, and technology grows, the question is raised as to whether it will remain an interdisciplinary field or establish its own discipline. U.S. researchers are active in the area and will continue to play a major role with major competition coming from the European Union (in the area of environmental aerosols) and Asia (Japan, in nanostructured materials).

Panel's Summary Assessment. The current U.S. position is at the "Forefront," and in the future, the United States will be "Maintaining" its relative position.

4.8.d Green Engineering

The cost to society for containing and eliminating the unintended consequences of chemical production can be greatly reduced by careful design of processes and products with a life-cycle analysis of their environmental, safety, and health effects. Examples of products that resulted in major societal and economic benefits when introduced, only to later have enormous health and environmental costs are DDT, freon, and tetraethyl lead. Such widely publicized problems have led to the recognition of the need for products that are environmentally acceptable. Additional major costs to society have resulted from the improper disposal of hazardous chemicals. The costs of the cleanup of contaminated sites over a 50-year period are projected to be $1 trillion dollars. Green engineering is the design of products and processes that will use natural resources and energy efficiently, and minimize harmful by-products and risk over the life cycle of the product. This clearly involves all disciplines, but especially the chemist, chemical engineer, environmental engineer, and toxicologist. The implementation of properly selected chemical reactions into product and process design with a life-cycle analysis to ensure that they meet environmental and health concerns involves other disciplines with chemical engineers playing a major role. "Development that meets the needs of the present without compromising the ability of future generations to meet their own needs," as sustainability is defined in the 1987 Brundlandt Report, is the goal. Translation of that goal into achievable engineering objectives is the challenge.

U.S. Position. Themes selected for the Virtual World Congress were sustainability, product engineering, technologies for a sustainable environment, and industrial ecology and life-cycle management. The United States contributed the majority (62%) of the speakers to the Virtual World Congress, with the European Union contributing most (34%) of the balance. Chemical engineers contributed a large majority (90%) of the speakers to process and product development, but their contribution dropped to 55% for the areas of life-cycle analysis and sustainability.

The publications for process and product development are in the mainstream chemical engineering journals and have been analyzed elsewhere in this report. The United States maintains a strong publication record in these journals, since green engineering cannot be easily viewed apart from broader chemical engineering activities. Publications in the *Journal of Environmental Engineering* are more focused on green engineering; contributions by U.S. authors represented 61% of the total; and contributions by chemical engineers exceeded those by the other peer disciplines. Designing green products requires a quantification of risk. The United States, with its preeminence in bioengineering and the complementary tools for risk assessment, is in a strong position here. U.S. authors contributed the largest percentage of the papers in *Environmental Toxicology and Chemistry*, although the percent contribution declined from 60% in 1997 to 44% in 2005, a consequence of significant increases in contributions from the European Union and Asia. The areas of the life-cycle analysis and industrial ecology are more broadly based with strong economic and sociological components. Publication rates are growing with more specialized journals being established in the past decade: *International Journal of Life Cycle Analysis* (1996), *Journal of Industrial Ecology* (1997), *Green Chemistry* (1999), and *Clean Technologies and Environmental Policies* (2002). The U.S. contributions to these journals are strong, but the European Union presence is dominant in those journals sponsored by European organizations. The United States is poorly represented in the 10 most-cited papers in *Green Chemistry* and the *International Journal of Life Cycle Analysis*. The chemical engineering contributions to life-cycle analysis and industrial ecology is smaller than those to product and process development, given the multifaceted dimensions of these disciplines.

Relative Strengths and Weaknesses. It takes well-publicized incidents such as DDT and Love Canal to energize public interest in environmental problems. The response to these problems has resulted in major reductions in the releases of persistent bioaccumulative toxins (PBTs), and major progress has been made in the cleanup of hazardous waste sites. The cost to society of these problems, both environmental and economic, have motivated the move to green chemistry and sustainability, not so much driven by the public but by the Environmental Protection Agency and industry. The Virtual World Congress and publications show a strong U.S. leadership in these areas, with competition coming mainly from the European Union, in part because of strong governmental and industry support for innovations in these areas. The United States has made significant contributions to the life-cycle assessment method of environmental impact accounting of products and processes, sustainability metrics, green chemistry themes and benign syntheses (e.g., supercritical CO_2, ionic liquids, and microwave), various design tools for pollution prevention approaches, and significant advances in cleaner production practices in industry. It is among the world leaders in energy intensity reduction and the development of market-based methods for alternative energy development and pollution prevention (P2) tools and methods.

The evolution of the response from drivers based on command and control (compliance based of reducing end of pipe emissions) to a more holistic sustainable development with constraints based on health, safety, and environmental (HS&E) considerations is directly related to the core skills of chemical engineers. The risk is that such skills as process development and design are threatened by the reduction of the corresponding research activities.

U.S. efforts have been supported by Environmental Protection Agency grants, programs complemented by the leadership of major chemical companies, which are showing that green chemistry and sustainable development are good business and are promoting such activities through the AIChE Centers for Waste Reduction Technology (CWRT) and Sustainable Technology Practices (CSTP). The commercial value of good environmental management has already been reflected in increased stock value for those corporations with superior management of environmental issues. While major progress has been made in reducing the environmental footprint of individual corporations, the establishment of the broader policy and economic framework that will lead to sustainable development is still in an evolutionary stage. Chemical engineers will be called to play a pivotal role, but the requisite skills are gradually deteriorating, thus threatening the success of the proposed enterprise.

Future Prospects. One of the major developments during the past 10 years has been the introduction of anticipatory approaches to pollution prevention through life-cycle analysis, product engineering, and new chemical synthesis routes. When coupled with improved life-cycle assessment methods of environmental impact accounting of products and processes, one has the essential framework for progress in this area. The integration of environmental impact assessment software into widely used process simulation, design, and optimization software offers the enabling tools. U.S. researchers have taken a healthy leadership position and have been involved in all of these efforts, but the European Union has taken a more decisive position.

One can expect a continued strong leadership position by U.S. researchers in the development of technologies for reducing the environmental footprint of chemicals in the environment through their life cycle, supported by the Environmental Protection Agency and industry. The challenge of sustainability will be more difficult to solve given its social, political, and economic dimensions, which need to be addressed worldwide. Of particular concern to the chemical engineer will be the availability of raw materials, which deplete over time, particularly natural gas and petroleum.

Of particular importance for the future are the following areas, which will require continuation or initiation of properly supported research efforts: Risk assessment of nanotech products; pollution prevention through nanotechnology; biotechnology as an enabling technology for green products and processes; and life-cycle impact from the development of alternative energy sources.

Panel's Summary Assessment. The current U.S. position is "Among World Leaders," and in the future, the United States will be "Gaining or Extending" its relative position.

4.9 AREA-9: PROCESS SYSTEMS DEVELOPMENT AND ENGINEERING

This area has always formed a key component of the core of chemical engineering, being concerned with concepts, tools, and techniques for the design, development, and exploitation of process systems in the broadest sense. Traditionally, the focus of activities has been on the design and operation of manufacturing systems for chemical products (chemical plants) based on traditional manufacturing components (unit operations), but there has been increasing interest in the development and exploitation of products, as well as in the development of novel manufacturing concepts (e.g., process intensification, or micromanufacturing). As in all research involving methodological developments, the application of new techniques to challenging practical problems is a key part of research in this area.

Four distinct subareas of research activities make up the research scope of this area:

- process development and design
- dynamics, control, operational optimization
- safety and operability of chemical plants
- computational tools and information technology

The mainstream chemical engineering journals, *AIChE Journal*, *I&EC Research*, and to lesser extent, *Chemical Engineering Science*, *Chemical Engineering Research and Design*, and the *Canadian Journal of Chemical Engineering*, along with the area-specific journals, *Computers and Chemical Engineering* and the *Journal of Process Control* have been the primary depositories of research contributions by chemical engineers in this area worldwide. A series of other subarea-specific journals have attracted a smaller number of very influential publications by chemical engineers and will be discussed later in this section.

Analysis of publications in the first three journals indicates that the ratio of U.S.-papers per non-U.S. papers has been reduced by roughly 50% during the last 5 years, as a result of rapid growth in research activity and output, primarily in Asia.

Analysis of the publications in *Computers and Chemical Engineering*, a popular journal in which to publish methodological contributions on theory, tools and techniques, and applications, is very instructive for the trends of chemical engineering research in process systems engineering in various geographic regions. Table 4.36 shows that while the number of U.S.-originated papers has grown by about 60% from the 1990-1995 to the 2000-2006 period, the relative percentage has remained roughly the same (37% in 1990-1994, 34% in 2000-2006). The corresponding percentage for European Union papers has been reduced, and Asian contributions have increased significantly (from 7% to 21%), a phenomenon which is completely in line with the rates of industrial investments in commodity plants observed in China and India during the past 10 years. Looking at the years 2003 to 2005, the total number of articles published in the journal (which obviously covers a broader field than only process design methodologies) grew from 129 in 2003 to 213 in 2005, a growth rate of 65%. U.S. contributions also grew from 51 to 71 (39% growth). The corresponding growth rates for Europe, China, and India were 81%, 114%, and 80%, respectively. For the latter two countries, the absolute numbers are currently small, but European contributions were of a similar scale to those from the United States in 2005, having been significantly lower in 2003 (see Table 4.37 below).

TABLE 4.36 Origin of Publications in *Computers and Chemical Engineering*

	1990-1994		1995-1999		2000-2006	
		%		%		%
Total No. of Papers	679		1,338		1,218	
United States	254	37	364	27	413	34
EU	238	35	421	31	319	26
Asia	47	7	160	12	253	21
Canada	19	3	35	3	68	6
S. America	31	5	90	7	170	14

TABLE 4.37 Geographic Distribution of Origin of Papers Published in *Computers and Chemical Engineering* in Recent Years

	United States	EU	China and India
2003	51	36	12
2004	64	60	17
2005	71	65	24

In terms of quality and impact, the U.S.-originated papers held a very commanding lead in the list of the 30 most-cited papers for the three periods, 1990-1994, 1995-1999, and 2000-2006, as shown in Table 4.38. However, one should not overlook the fact that with increased levels of research activity, the overall quality improves. Indeed, in 1990-1994, Asian and South American countries had no representation in the list of the 30 most-cited papers. In the period 2000-2006 the number is seven.

4.9.a Process Development and Design

Included in this subarea are methodologies, tools, and techniques to aid engineers in the synthesis, development, and design of new manufacturing systems (e.g., single plants, supply chains). Systematic and integrated handling of raw materials pretreatment, synthesis of reactor configurations, of separation trains, and of energy management systems is at the heart of rational process development. In addition, for batch processes the early integration of synthetic chemists and chemical engineers is essential for the early evaluation of alternative synthetic routes and the selection of the most promising processing schemes from an economic and environmental point of view. Rational strategies for process scale-up remain a subject of importance.

Research on novel manufacturing concepts (e.g., process intensification, miniaturization) also falls under this heading and involves skill sets brought forth by a variety of skilled chemical engineers. With the emphasis on molecular-level understanding that characterizes current trends in chemical engineering research, systematic approaches to process development are being explored for micro- and nanoscale processes.

U.S. Position. The number of experts in this subarea was 16, with 10 (63%) from the United States. Of the speakers, 57% of nominations were for U.S.-

TABLE 4.38 Distribution of the 30 Most-Cited Papers in *Computers and Chemical Engineering*

	1990-1994	1995-1999	2000-2006
United States	28	21	20
EU	2	7	3
Asia	0	0	3
Canada	0	1	0
S. America	0	0	4

based researchers when duplication of names was allowed and 59% when duplication was disallowed. These results indicate that the United States holds a leadership position in this subarea, although the strength of that position is not as pronounced as in other areas considered in this study.

Given the breadth of the subarea, and the fact that many contributions are published in "generalist," mainstream chemical engineering journals (*AIChE Journal, Chemical Engineering Science*, and *Industrial and Engineering Chemistry Research*), it is difficult to draw reliable conclusions for this topic from an analysis of publications. A limited examination of publications in these three journals indicated that the number of U.S.-papers per non-U.S. paper has decreased by about 50%, in line with trends we have seen for these journals in other subareas.

Relative Strengths and Weaknesses. Within the broader scope of process systems engineering, U.S. chemical engineering research activities in process development and design took an early leadership position since the pioneering activities on process synthesis in the late 1960s to early 1970s. U.S. leadership strengthened with the entry of many young U.S. chemical engineering researchers into the field and the parallel deployment of their ideas into industrial practice. During the following 20 years process synthesis research was introduced in the undergraduate curricula of U.S. chemical engineering, and its reach encompassed most countries of the world. As a result of all these developments, U.S. academic and industrial activities in process synthesis led to a significant competitive advantage, especially in continuous processes, and the introduction to the market place of a series of computer-aided tools by software and engineering services companies (founded and managed by chemical engineers). Analogous activities for batch processes started in the early 1980s and have led to similar systematization of process development for the pharmaceutical and fine chemicals industries. Relevant computer-aided tools have also been developed and are being marketed, primarily by U.S.-based software and engineering services companies.

However, during the past 15 years we have witnessed a gradual deterioration in the funding and the level of research activities associated with process synthesis, for continuous and, to a lesser extent, for batch processes. A number of research laboratories and centers have closed or have reduced the level of their research activity significantly. The primary reason has been the shift in the strategic plans of major commodity chemical companies, and the collateral effect on federal funding supporting such activities. Consequently, although a number of people with high skills in process synthesis and design are presently working in U.S. chemical companies, the number of new graduates with research experience in this area has dropped dramatically. While no one expects that chemical companies will start building new petrochemical plants in the United States, process synthesis and design are essential skills for the development of new generations of petrochemical plants, built by U.S. companies in Asia and the Middle East; cellulose-based ethanol plants; multipurpose batch plants for the pharmaceutical and fine chemicals industries; and plants for the manufacturing of a broad variety of functional materials. All of these areas are arguably of significant interest to U.S. chemical companies for the needs of the U.S. economy. However, the skilled human resources who would enable such a resurgence may not be available if the level of research activities in process synthesis and design continues to drop.

Future Prospects. Some of the most significant advances in process development and design during the past 10 years are the following: industrial implementation of systematic process synthesis methodologies and algorithmic procedures for continuous and batch plants; widespread implementation of residual curve maps for the design of distillation separations; engineering of integrated process networks (e.g., reaction, energy, mass, and water); process intensification (e.g., microplants, modular plants); and integrated process design and control. U.S. researchers have been leading contributors in all of these developments. Europe is very strong, while Asian contributions have dealt primarily with specific applications.

In the near future, research is expected to focus on processes with lower levels of energy consumption, high-throughput synthesis of pharmaceuticals and fine chemicals with parallel consideration of process development, process intensification (e.g., plants on a chip), green production routes with parallel process development, and design of novel hybrid unit operations integrating reactions and separations.

One of the most interesting developments during the past 10 years is the emergence of systematic product design as a subject of chemical engineering research. Given the current trends of an increasingly product-centric chemical industry, this interest will continue and will become more closely integrated with the design of the process on which the manufacturing of the product is based.

U.S. researchers are well positioned to address these needs, provided that sufficient support becomes available.

Panel's Summary Assessment. The current U.S. position is "Among World Leaders," and although in the future this position is expected to weaken, due to uncertainties in funding, the United States will remain "Among World Leaders."

4.9.b Dynamics, Control, and Operational Optimization

This subarea is concerned with research in support of achieving operational excellence and covers process control; optimization of various aspects of operational performance (online optimization of steady-state and transitional operations, including startup and shutdown, performance for continuously operated plants, trajectory optimization for batch plants); and scheduling and supply-chain management. Monitoring and control of polymer processes, microelectronic fabrication, biological processes, microchemical processes, electrochemical processes, as well as planning, scheduling, and supply-chain management and dynamic simulation and optimization, are a few of the current research interests in chemical engineering for this subarea. The underlying numerical methodologies and computer-aided tools of analysis and design for control systems, optimization of large-scale integrated plants and/or supply chains of plants, and simulation of nonlinear steady-state or dynamic processes are within the scope of several disciplines (e.g., for control, electrical, mechanical, and aerospace engineering; for optimization, operations research; for dynamic simulation, applied math). However, chemical engineers have been the unique enablers of the application of these methods and tools in the chemical industry at large, and have led several breakthrough developments, notably the introduction of advanced model-based control in chemical processes and methods for global mixed-integer optimization. Furthermore, theoretical and methodological contributions from chemical engineers in, for example, control and optimization, have had broad impact in other disciplines.

U.S. Position. U.S. representation among the experts for this subarea was 8 from 12, i.e., 67%. Out of 212 total nominations, 122 were for U.S.-based speakers (58%). These results indicate a leadership position in this subarea for the United States.

AIChE Journal, *I&EC Research*, and *Computers and Chemical Engineering* have attracted a sizeable fraction of process control and optimization papers by chemical engineers worldwide. A close analysis of these papers indicates that the contributions in control and optimization follow similar lines as those described above for all publications in these journals. The *Journal of Process Control* is a popular medium of control-related publications by chemical engineers. Table 4.39 shows a comparison of publication rates from the 1995-1999 and 2000-2006 periods by different geographical regions. The figures indicate a marginal, though possibly insignificant, decline in the U.S. share of contributions between the two 5-year periods, with strong growth (in share and numbers) from the European Union and Canada. Data for the past 5 years (see Table 4.40) indicate a stronger decline in the percentage of U.S.-originated publications.

TABLE 4.39 Geographic Distribution of the Origin of Papers Published in *Journal of Process Control*

	1995-1999		2000-2006	
		%		%
Total No. of Papers	217		433	
United States	68	31	125	29
EU	46	21	119	27
Asia	52	24	111	26
Canada	22	10	67	15
S. America	8	4	21	5

TABLE 4.40 Papers Published in *Journal of Process Control* in Recent Years

	2000	2001	2002	2003	2004	2005	2006 (part)
Total Number of Papers	53	61	62	69	65	92	31
No. of U.S. Papers	23	23	18	14	18	21	8
U.S. Papers	43	38	29	20	28	23	26

Whilst the number of total articles published in the journal since 2000 has grown significantly (from 53 in 2000 to 92 in 2005), the number of articles originating from the United States has stayed roughly constant at best, and as a result, the U.S. share of contributions has fallen significantly since 2000.

An analysis based on the 30 most-cited papers from the journal is shown in Table 4.41. The results reveal that the United States and the European Union maintain a strong lead based on this criterion, although it is perhaps too early for these data to be affected by the recent significant decline in the U.S. share of contributions.

TABLE 4.41 Distribution of the 30 Most-Cited Papers in *Journal of Process Control*

	1995-1999	2000-2006
United States	11	16
EU	11	10
Asia	2	2
Canada	5	2
S. America	1	0

The process control community in chemical engineering is part of the broader automatic control community, and we were interested to seek information on the position of chemical engineers, and of U.S. chemical engineers in particular, within that broader grouping.

Popular journals for automatic control researchers are *Automatica* and *IEEE Transactions on Automatic Control*. An analysis of the chemical engineering contributions to those journals is shown in Table 4.42.

The proportion of chemical engineering contributions to these generalist journals is clearly low, and there is little evidence of growth in the past 15 years. (Absolute numbers of contributions from chemical engineers have grown, but at a rate in line with overall growth in contributions from all disciplines.) A very striking feature of the chemical engineering contributions is the dominant position of U.S. authors. This is illustrated in Table 4.43 where percentages of contributions featuring chemical engineering authors from various geographical regions to *Automatica* are presented.

In the area of optimization, most of the contributions by chemical engineers are published in the journal *Computers and Chemical Engineering*. The relative contributions by U.S. and non-U.S. authors follow similar trends as those discussed earlier for the journal at large.

In the area of optimization, chemical engineers have been publishing in a variety of specialized journals, like the *Journal of Optimization Theory and Applications, Mathematical Programming, INFORMS Journal on Computing*, and others (see Table 4.44). The numbers of papers and

TABLE 4.42 Percentages of Papers Featuring Chemical Engineering Authors Published in *Automatica* and *IEEE Transactions on Automatic Control* by Time Period

	1990-1994	1995-1999	2000-2006
Automatica	3.3	4.1	3.4
IEEE Trans. Automatic Control	1.2	1.2	0.7

TABLE 4.43 Percentages of Papers in *Automatica* with Chemical Engineering Authors by Geographical Region (Papers with authors from more than one region have been counted for each region featured.)

	1990-1994	1995-1999	2000-2006
United States	18.2	46.0	55.3
EU	45.5	13.5	19.2
Asia	18.2	18.9	25.5
Canada	18.2	29.7	17.0

TABLE 4.44 Chemical Engineering Contributions to the Optimization Literature (2000-2006 August)

	No. of Chem. Eng. Papers	Total No. of Papers
Mathematical Programming	5	579
J. Optimization Theory and Applications	4	834
J. Global Optimization	19	550
Annals of Operations Research	7	799
INFORMS J. on Computing	1	210
Optimization and Engineering	8	50
SIAM J. on Optimization	4	420
Computational Optimization and Applications	5	314
SIAM J. on Scientific Computing	11	739

corresponding percentages are small: About 1% to 3.5% were contributed almost exclusively by a small number of U.S. academic researchers, leading to very large per capita numbers of papers. The percent contributions are quite healthy, given the extensive interdisciplinarity of these journals, and the quality of the chemical engineering contributions is usually high, set by a very competitive interdisciplinary group of researchers.

Relative Strengths and Weaknesses. As with process development and design, discussed in the previous paragraph, the U.S. chemical engineering community took an early lead in theoretical and applied process control and optimization activities in the mid 1960s. It was not until the mid to late 1970s that major breakthroughs in process control were introduced in the operation of large-scale chemical plants. The subsequent growth of industrially relevant and effective process control was rapid. The number of research groups around the country increased significantly, and the population of graduate students with education and skills in process dynamics and control expanded rapidly. During the 20-year period 1975-1995, process control research expanded to include control synthesis for complete chemical plants, integration of regulation and operational optimization, design of multivariable optimal regulators for fairly large systems, and fairly sophisticated diagnostic methodologies for the early detection of process faults, and promised to materialize the concept of an "operator-less" plant. In addition, advances in dynamic simulation opened the door to complex nonlinear control systems, and the expansion of optimization capabilities allowed the optimal planning, scheduling, and control of a large number of batch operations. It should be noted that chemical engineers have contributed substantially more than other engineering disciplines in advancing the theory and industrial practice of interdisciplinary areas, such as nonlinear programming, optimization with integer and continuous variables, and global optimization.

All of these achievements are presently at risk. For the past 10 years we have witnessed the gradual reduction in the level of research activities in process control and optimization. Federal funding and industrial support for such research have been reduced. Academic researchers in process systems engineering have turned their attention to problems for which they can secure funding. While such reorientation is healthy in many respects, it has undermined the broad-impact breakthroughs that came with earlier research, and while it helps maintain certain low numbers of graduates skilled in process control and optimization, it has undermined the morale of U.S. researchers in this area.

Ensuring that adequately trained human resources are available in sufficient numbers to ensure success in the new challenges, analogous to those described in the previous section on process development and design, is the most critical issue for this subarea.

Future Prospects. The rapid growth in the number of model-predictive control (MPC) systems installed in chemical plants and their integration with operational optimization algorithms in real time are two of the significant developments during the past 10 years. In addition, very effective optimization algorithms for large-scale and nonlinear supply-chain problems have resulted in significant shifts of industrial practices. U.S. academic and industrial researchers and engineers have driven most of the theory and applications development of MPC in the chemical industry, and the principal contributions in large-scale optimization theory have come from the United States. The European Union is very strong in all the subject matters of this subarea, and Asian researchers have focused primarily on applications.

Research towards the development of model-predictive control systems, which monitor, diagnose, and adapt their performance, and parametric programming for process control are well on their way for industrial implementation, but still need support for their successful completion. Industrial needs for commodity chemical plants require further development of online and large-scale dynamic process optimization algorithms with the ability to monitor, diagnose, and adapt their search and performance. Control of multiscale and distributed processes, and model-predictive control and operational optimization of nonlinear and hybrid processes will become more prominent in the future, especially for materials- and device-manufacturing processes with quality specifications at small scales and many discrete operations. Online process monitoring for product quality assessment will also attract more interest for such manufacturing systems.

Panel's Summary Assessment. The current U.S. position is at the "Forefront," and although in the future this position is expected to weaken, due to uncertainties in funding, the United States will remain "Among World Leaders."

4.9.c Plant Operability and Safety

This subarea involves research into the identification and mitigation of hazards associated with the operation of manufacturing facilities, as well as all practical engineering considerations associated with safe, smooth, flexible, resilient, and robust operability of such facilities.

U.S. Position. For the Virtual World Congress 11 experts were consulted with 10, i.e., 91%, of them being from the United States. U.S.-based speakers represented 77% of nominations (137 out of a total of 179) when duplications were allowed. This number dropped to 69% (70 out of 102), when duplications were disallowed. These results indicate a clear leadership position in this subarea for the United States.

Key journals in this area are published by national chemical engineering professional bodies: *Process Safety Progress* is published by the American Institute of Chemical Engineers, and *Process Safety and Environmental Protection* by the Institution of Chemical Engineers based in the United Kingdom. The proportions of U.S. papers published in these two journals reflect their geographical origins: in 2005, 77% of the papers published in *Process Safety Progress* featured U.S.-based authors; for *Process Safety and Environmental Protection* the corresponding figure was as low as 10%. It is difficult to argue that these results provide confirmatory evidence of U.S. leadership for this area. Indeed, the higher proportion of non-U.S. contributions in *Process Safety Progress* than of U.S. contributions in *Process Safety and Environmental Protection* might be argued to show relative weakness of U.S. research internationally in this area.

Future Prospects. Large-scale data reconciliation, process monitoring and fault detection for continuous commodity plants, and advanced systematic methods for the identification of hazards and safety analysis have been the most significant advances in the past 10 years. Efforts along these lines for advanced methods will continue, as the implementation of new technologies requires shifts in operating procedures and management of operations. The Panel expects that the scope of traditional concerns on safety will expand to include the evolving and more stringent constraints on environmental impact. This is a fertile area of future research, since it leads to an integrated approach in process conceptualization, process design and process safety, and operability and control. Computer-aided systems for integrated hazards-safety-risk assessments will also become necessary, and the need will increase for new sensor designs, data visualization, and image processing and analysis.

Panel's Summary Assessment: The current U.S. position is "Among World Leaders," and in the future is expected to remain "Among World Leaders."

4.9.d Computational Tools and Information Technology

Mathematical and computational modeling is an underpinning technology supporting research in many areas of chemical engineering. This subarea includes research in methods and tools for the modeling and simulation of process systems. Dynamic simulation of nonlinear systems (hybrid or not), dynamic pattern formation, modeling and analysis of multiscale systems, complexity theory and modeling/analysis of complex systems, as well as knowledge extraction from operating data, large-scale information processing for enhanced performance, security, and environmental impact, knowledge management and organizational learning, and aspects of an emerging cyber infrastructure, are a few of the issues attracting current research interests. The computational challenges associated with resolving the complex mathematical and computational problems that arise are often significant. As a result, chemical engineering researchers are making important contributions to the fundamentals of computation, through the development of concepts, methods and algorithms to handle complex process systems problems. Other important areas of computing, such as decision support and the organization, retrieval, and interpretation of large complex datasets, are also included in this subarea.

U.S. Position. The number of experts for the Virtual World Congress in this subarea was seven, with five (71%) from the United States. Of the speakers, 63% nominations were for U.S.-based researchers. These results indicate that the United States holds a leadership position in this subarea.

Computers and Chemical Engineering is a popular journal in which to publish contributions on the topics of this subarea. Analysis of the papers indicated that the general trend observed for the journal at large (see above) hold true for the contributions in this subarea. Chemical engineering researchers contribute little to interdisciplinary journals in this subarea, such as *SIAM Journal on Scientific Computing* (1.4%), *International Journal on Numerical Methods in Engineering* (0.7%), *International Journal on Bifurcation and Chaos* (0.4%), and others with smaller fractional contributions.

Relative Strengths and Weaknesses. Advanced methods and computer-aided tools for modeling, analysis, and simulation of processing systems and sophisticated information management systems form the underpinnings of all process systems engineering activities and have a critical effect on the deployment of all systems engineering tasks, such as product and process development and design, process control, supply-chain management and optimal planning and scheduling of process operations, and process monitoring and fault detection. These technologies along with the infrastructures that allow the coordinated aggregation and interaction of software, hardware, and human researchers and engineers, have a critical effect on the creativity and productivity of the chemical industry and have led chemical operations to unprecedented levels of operational efficiency.

Over the last 45 years, a large and vibrant community of U.S. (and United Kingdom) academic researchers and industrial practitioners established this subarea as a pole of significant attraction for talented young people. The results of their work fueled the generation of a series of commercial products with global reach, which have substantially increased the effectiveness and productivity of chemical engineers. The highly sophisticated process design and engineering allowed U.S. chemical companies to lead the competition in process licensing around the world.

However, today the systems engineering infrastructure (human and technological) of the U.S. chemical industry is at risk of losing its preeminence and competitive advantage. The number of active researchers in this subarea has decreased significantly during the past 10 years. The primary reason for this decrease has been a significant reduction in available funding for research in this subarea. The corresponding number of research groups and graduating PhD students is very low as well. Research in the design and deployment of a modern "cyber infrastructure" is not taking place, threatening a deterioration in the productivity and competitiveness of new chemical processes (independently of the geographic location) and the creativity and effectiveness of the industrial research enterprise in health-care products (pharmaceuticals, diagnostic products), fine chemicals, functional materials, biomass-based fuels, and new energy devices.

Future Prospects. The establishment of the CAPE-OPEN standards and the opening of the path for the design of plug-and-play software in process systems engineering is one of the most interesting developments during the past 10 years. In addition, effective algorithmic approaches have been developed for modeling, simulation, and optimization of continuous, discrete-event, hybrid, and multiscale dynamic processes. Simulation, design, and optimization under uncertainty, very effective global optimization algorithms, and the expanding use of Monte Carlo simulators, along with the advances mentioned above, have enhanced the abilities of chemical engineering researchers in many subareas by offering the tools they need for materials and peptides design, metabolic engineering, green engineering, combustion, kinetics and reaction engineering, and others.

The single most important development for the future may be the systematic design, deployment, and utilization of large-scale cyber infrastructures. Such systems, which will provide transparent integration of algorithmic procedures, databases, experimental equipment, and human researchers, may have far-reaching effects on the creativity and efficiency of chemical engineering research in all subareas. A subset of the possibilities includes biocatalysis and protein engineering; cellular and metabolic engineering; engineering of green products and processes; design of new materials; design and simulation of self-assembled systems; integrated product and process design; and integration of chemical production routes with process conceptualization and design, process safety, operability, and control.

The Panel believes that the need for decision-making, computer-aided tools that support efforts in the area of sustainability (e.g., dealing with uncertainty, multiple objectives, and complexity) will become more prominent in the future. The pressure for continuous improvements in the following areas of computational tools and information systems will remain strong: global optimization; multiscale and multi-agent process systems engineering; problem-specific mixed-integer optimization approaches; complexity and engineering design; and tools for visualization of data and operations.

Panel's Summary Assessment. The current U.S. position is "Among the Leaders," and in the future is expected to remain "Among World Leaders."

4.10 SUMMARY

Based on the analysis of data regarding the composition of the Virtual World Congress, publications and citations, patents, recognition of individual researchers through prizes and awards, and prevailing trends, the Panel compiled an overall assessment for each subarea in terms of the following two indices:

- Current Position of U.S. Research in Chemical Engineering
- Expected Future Position of U.S. Research in Chemical Engineering

Table 4.45 summarizes the Panel's assessment of the Current and Expected Future Positions of U.S. Chemical Engineering Research in all

TABLE 4.45 Assessment of Current and Future Positions for U.S.
Chemical Engineering Research
(X = current position; grey circle = future position)

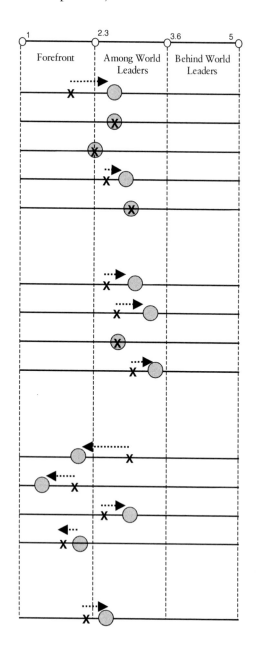

Materials

Polymers

Inorganic and Ceramic Materials

Composites

Nanostructured Materials

Biomedical Products and Biomaterials

Drug Targeting and Delivery

Biomaterials

Materials for Cell & Tissue Engineering

Energy

Fossil Energy Extraction and Processing

Fossil Fuel Utilization

Non-Fossil Energy

Environmental Impact and Management

Air Pollution

Water Pollution

Aerosol Science and Technology

Green Engineering

Process Systems Development and Engineering

Process Development and Design

Dynamics, Control, and
        Operational Optimization
Safety and Operability
        of Chemical Plants
Computational Tools
        and Information Technology

subareas alongside the expected future trends. The major conclusions are as follows:

Conclusion 1: U.S. chemical engineering research is strong and at the "Forefront" or "Among World Leaders" in all subareas of chemical engineering. It is expected to remain so in the future.

Conclusion 2: U.S. research is particularly strong in fundamental engineering science across the spectrum of scales: from macroscopic to molecular. In these areas of research, the primary competition in terms of quality and impact comes from other disciplines rather than from chemical engineers from other countries. However, recent trends of increasing levels of applications-oriented research with a parallel decrease in the levels of basic research will continue and may undermine the historical strength and preeminence of U.S. chemical engineering.

Conclusion 3: In the core areas of chemical engineering research, the level of output from Asian and European Union countries has increased significantly during the past 10 years, but the United States maintains a strong leadership position in terms of quality and impact.

Conclusion 4: In the following subareas of chemical engineering research, the United States will be "Gaining or Extending" its current relative position: biocatalysis and protein engineering; cellular and metabolic engineering; systems, computational, and synthetic biology; nanostructured materials; fossil energy extraction and processing; non-fossil energy; and green engineering.

Conclusion 5: The Panel has recognized that funding policies (government and industrial) may put at risk the U.S. position in the following subareas of chemical engineering research: transport processes; separations; catalysis; kinetics and reaction engineering; electrochemical processes; bioprocess engineering; molecular and interfacial science and engineering; inorganic and ceramic materials; composites; fossil fuel utilization; process development and design, and dynamics, control, and operational optimization.

Conclusion 6: The degree of interdisciplinarity varies from subarea to subarea but is significant in all areas of chemical engineering research and in recent years has been growing. Therefore, the future competitiveness of U.S. chemical engineering research must be benchmarked against a broader spectrum of disciplinary contributions.

Conclusion 7: Trends in research funding policies will continue to reduce chemical engineering's dynamic range, strengthening its molecular orientation in bio- and materials-related activities at the expense of research in macroscopic processes.

5

Key Factors Influencing Leadership

In the context of this report, research leadership in chemical engineering has been measured by various factors such as numbers and citations of journal articles and a Virtual World Congress conducted by the panel members. This leadership is influenced by a multitude of factors that are largely the result of national governance, structural and support polices, and overall available resources of each country in the world. As done previously,[1] the panel focused on four key factors that influence the international leadership status of the U.S. chemical engineering research:

- *Innovation*: Investment and technology development mechanisms that facilitate introduction of chemical science and technology into the marketplace.
- *Major facilities, centers, and instrumentation*: The physical infrastructure and materiel for conducting chemical engineering research.
- *Human resources*: The national capacity of chemical engineering students and degree holders.
- *Funding*: Financial support for conducting chemical engineering research.

---

[1]National Research Council, *Experiments in International Benchmarking of US Research Fields*, National Academy Press, Washington, D.C., 2000.

5.1 INNOVATION

A key factor influencing leadership in chemical engineering is how rapidly and easily new ideas can be tested, developed, and extended into the U.S. economy as well as the global marketplace. This process by which research ideas are developed and funded in the United States has been defined as our "innovation system." The U.S. innovation system, like that in other countries, is characterized by a set of unique attributes. Some of the factors that influence the U.S. innovation process for the field of chemical engineering are discussed below.

5.1.a A Strong U.S. Industrial Sector

Leadership in chemical engineering research in the United States over the years has been strongly linked with the development of the U.S. chemicals industry. According to Landau and Arora,[2] "the rise of the research university in science and engineering gave a strong boost to the American chemical industry" particularly in the early part of the 20th century. And this relationship has been a vital part of the success of the United States as a nation. Landau and Arora further point out that the U.S. chemicals industry: (1) "was the first science-based, high-technology industry"; (2) "has generated technological innovations for other industries, such as automobiles, rubber, textiles . . ."; and (3) "is a U.S. success story."

At the same time, the U.S. chemical manufacturing industry is not what it used to be. Once a major net exporter, the U.S. chemical industry is now essentially a net importer (trade went negative in 2000-2001).[3] Some feel that today the U.S. chemical industry is in fact fundamentally disadvantaged relative to the rest of the world because of its dependence on oil and natural gas for raw materials, which have become less abundant and much more costly than they used to be. The chemical industry consumes only 5% of the total production of oil and natural gas, while the majority is used in transportation, residential, and other industrial requirements such as energy generation; and the cost of natural gas is 2 to 10 times higher than anywhere else in the world. This is greatly influencing investment for new plants, jobs, and even research outside the United States.[4]

---

[2]R. Landau and A. Arora, "The dynamics of long term growth: Gaining and losing advantage in the chemical industry," Pp. 17-43 in *U.S. Industry in 2000: Studies in Competitive Performance*, D. C. Mowery, ed., National Academy Press, Washington, D.C., 1999.

[3]W. J. Storck, "UNITED STATES: Last year was kind to the U.S. chemical industry; 2005 should provide further growth," *ChemicalEngineering News* 83 (2):16-18.

[4]M. Arndt, "No longer the lab of the world." *Business Week*, May 2, 2005.

5.1.b A Variety of Funding Opportunities

Another key attribute of the U.S. innovation system is the existence of a multitude of funding options—from largely government-supported academic research to entrepreneurial work supported by small and large companies. This variety of sources, with different emphases, creates a spectrum of opportunities for chemical engineering research.

Industry

As we will discuss later, this sector is the largest supporter of R&D. Individual companies may operate their own R&D labs as well as provide funds for academic topical/strategic research.

Federal Government

The National Science Foundation (NSF) Engineering Research Center (ERC) and Science and Technology Center (STC) models are intended to spur innovation. While NSF mainly supports academic research, it seeks to foster successful links between academe and industry with programs such as Grant Opportunities for Academic Liaisons with Industry (GOALI) and Integrative Graduate Education and Research Traineeship (IGERT). NSF also has more directed collaborative research and education programs in the area of nanoscale science and engineering, such as Nanoscale Interdisciplinary Research Teams (NIRT), the Nanoscale Exploratory Research (NER), and Nanoscale Science and Engineering Centers (NSEC). Other federal mission agencies (Department of Defense, Department of Energy, National Institutes of Health, and the National Institute for Standards and Technology, also fund a great deal of physical science and engineering.

The Small Business Administration (*http://www.sba.gov*) supports the agency-wide Small Business Innovative Research program (SBIR), which is a highly competitive program that encourages small businesses to explore their technological potential and provides the incentive to profit from its commercialization. Each year, 10 federal departments and agencies are required to reserve a portion of their R&D funds for awards to small business. The Small Business Technology Transfer program (STTR) is another important small business program that expands funding opportunities in the federal innovation research and development arena. Each year, just five federal departments and agencies (Department of Defense, Department of Energy, Department of Health and Human Services, National Aeronautics and Space Administration, and National Science Foundation) are required by STTR to reserve a portion of their R&D funds for awards to small business/nonprofit research institution partnerships.

State Initiatives

There have also been a growing number of state initiatives to foster innovation and stimulate economic growth:

- Pennsylvania Infrastructure Technology Alliance (*http://www.ices. cmu.edu/pita*) is a program that is designed to aid in the transfer of knowledge to provide economic benefit to the state of Pennsylvania.
- Texas Technology Initiative (*http://www.txti.org*) is a long-term economic development strategy designed to retain and attract advanced technology industries, coordinate advanced technology activities throughout the state, and accelerate commercialization from R&D to the marketplace to drive new business development in the state.
- New York State office of Science, Technology, and Academic Research (NYSTAR—*http://www.nystar.state.ny.us*) has a technology transfer innovation program (TTIP), which funds academic research that has a New York State industry partner that cost shares some of the work.

Universities

Many universities are now putting more funding towards supporting research, especially through centers that provide community outreach, span multiple universities, and even partner with industries. Examples include the following:

- The University of California solicits proposals for "UC Discovery Grants" in biotechnology to promote industry-university research partnerships. Biotechnology is one of five fields supported by UC Discovery Grants (i.e., biotechnology, communications and networking, digital media, electronics manufacturing and new materials, and life sciences information technology). UC Discovery Grants enhance the competitiveness of California businesses and the California economy by advancing innovation, R&D, and manufacturing, and by attracting new investments.
- Pennsylvania State University, Center for Glass Surfaces, Interfaces, and Coatings (Carlo G. Pantano)
- Lehigh University, Center for Optical Technologies (*http://www. lehigh.edu/optics*)

Private Foundations

There are many philanthropic organizations that help round out the support for chemical engineering R&D in the United States, such as:

- The Camille and Henry Dreyfus Foundation, Inc. (*http://www. dreyfus.org*)
- The Research Corporation (*http://www.rescorp.org*)
- The American Chemical Society Petroleum Research Fund
- Bill & Melinda Gates Foundation (*http://www.gatesfoundation. org/default.htm*)

Venture Capital

Chemical engineers are increasingly involved in small business startup companies that often seek out venture capital funding. This is especially the case for biotech, semiconductor, and medical device research applications. For example, a startup firm proposing a completely new, biological means of laying down thin films and carrying out other steps in electronics manufacturing secured financing worth more than $12 million from investors that included nanotechnology specialist Harris & Harris and In-Q-Tel, a venture capital group funded by the Central Intelligence Agency.[5]

5.1.c Cross-Sector Collaborations and Partnerships

Collaboration of university and industry researchers is another important aspect of the U.S. innovation system. Even though U.S. industry funds only about 10% of the research carried out in universities, the mobility of individuals between academic and industrial laboratories is especially vital in the transfer of new concepts and technology. In the past, many academics had significant industrial experience, where they interacted closely with industry in research and as consultants. Today, the majority of new faculty members come from academic labs where they have carried out postdoctoral research, such that the link to industry has been weakened. University faculty members also participate in the formation of high-tech companies. These relationships provide university researchers with an understanding of problems that are relevant to industry, and they provide a channel for the transfer of knowledge and new approaches developed in academia with funding from the federal government.

A good example of one industry-university-government collaboration is between the Chemical Engineering Department at University of Delaware, Rohm and Haas, Engelhard (now BASF), with funding from the Department of Energy. The program seeks to develop a major new manufacturing process that will use propane instead of propylene to manufacture acrylic acid. The novel technology, if adopted worldwide by acrylic acid and other propylene derivative manufacturers, could save up to 37 trillion BTUs per

---

[5] *http://pubs.acs.org/cen/news/83/i40/8340cambrios.html.*

year, eliminate 15 million pounds of environmental pollutants annually, and potentially save U.S. industry nearly $1.8 billion by the year 2020.

Such partnerships in general—whether between universities and industry or among companies—have become critical to improving the effectiveness with which industry commercializes research. However, many larger companies no longer carry out the level of exploratory research they once did, and U.S. universities can sometimes present significant barriers when it comes to intellectual property ownership. At the same time, other regions of the world that are presently accommodating in their licensing policies are increasingly moving toward the U.S. model of academic licensing.

5.1.d Strong Professional Societies

The American Institute for Chemical Engineering (AIChE) provides strong support for chemical engineering research in the United States as well as the world at large through the publishing of high-quality scholarly journals, holding annual meetings, and making connections between chemical engineers and the broader community. AIChE is a nonprofit professional association of more than 40,000 members that provides leadership in advancing the chemical engineering profession. Through its many programs and services, AIChE helps its members access and apply the latest and most accurate technical information; offers concise, targeted, award-winning technical publications; conducts annual conferences to promote information sharing and the advancement of the field; provides opportunities for its members to gain leadership experience and network with their peers in industry, academia, and government; and offers members attractive and affordable insurance programs. In addition, the American Chemical Society supports both chemistry and chemical engineering R&D efforts.

5.2 CENTERS, MAJOR FACILITIES, AND INSTRUMENTATION

Chemical engineering research is at the interface with many other disciplines, requiring specialized facilities (hardware, software) used by several other disciplines. Therefore the health and competitiveness of chemical engineering research depends on the health and availability of cutting-edge facilities at U.S. universities and national laboratories. The Office of Basic Energy Sciences at the Department of Energy[6] funds and operates several major facilities of relevance to chemical engineers that will be highlighted below: synchrotron radiation light sources, high-flux neutron sources, electron beam microcharacterization centers, nanoscale science research centers, and specialized single-purpose centers. There are also many

---

[6]*http://www.er.doe.gov/production/bes/BESfacilities.htm.*

National Science Foundation-funded centers and facilities, but these tend to be for used more heavily at the local university level—or with nearby universities. However, some of these centers do span multiple universities and provide an invaluable resource at the national level (some examples are included below). When available, important international facilities are included in the lists as well.

The types of facilities of interest to chemical engineering research fall into the following broad categories:

- materials synthesis and characterization facilities
- materials micro- and nanofabrication
- genetics, proteomics, and biological engineering
- fossil fuel utilization facilities (combustion centers)
- cyberinfrastructure (supercomputing)

5.2.a Materials Synthesis and Characterization Facilities

Synthesis and characterization of materials often requires high-energy light sources—such as synchrotron and neutron sources—or other specialized facilities that need a significant level of funding to operate and maintain. These are typically only available at national facilities, both here and abroad.

- Examples of important synchrotron sources include[7] Advanced Light Source (ALS), Advanced Photon Source (APS), National Synchrotron Light Source (NSLS), Stanford Synchrotron Radiation Laboratory (SSRL), Los Alamos Neutron Scattering Center, IPNS (Intense Pulsed Neutron Source) at Argonne and High Flux Isotope Reactor at Oak Ridge National Laboratory in the United States; Berliner Elektronenspeicherring-Gesellschaft für Synchrotronstrahlung (BESSY) in Germany; European Synchrotron Radiation Facility (ESRF) in France; INDUS 1/INDUS 2 in India; and National Synchrotron Radiation Research Center (NSRRC) in Taiwan.
- Examples of important neutron sources include[8] Spallation Neutron Source, Oak Ridge National Laboratory, and the University of Missouri Research Reactor Center in the United States; ISIS-Rutherford-Appleton Laboratories in the United Kingdom; and Hi-Flux Advanced Neutron Application Reactor in Korea.

---

[7]For a full list of worldwide synchrotron light sources, see *http://www.lightsources. org/cms/?pid=1000098*.

[8]For a full list of worldwide neutron sources, see the National Institute of Standards and Technology Center for Neutron Research at *http://www.ncnr.nist.gov/nsources.html*.

5.2.b Materials Micro- and Nanofabrication

Most research intensive universities are well equipped with conventional micro- and nanofabrication techniques such as thin-film deposition (e.g. chemical vapor deposition, physical vapor deposition), lithography, chemical etching, and electrodeposition, as well as characterization techniques such as electron microscopy, electron and X-ray diffraction, and probe microscopy that are used routinely to characterize small structures, small volumes, and thin films. However, the ability to characterize extremely small nanostructures or to tailor materials at an atomic level requires much more specialized equipment.

The Department of Energy is now in the process of opening five Nanoscale Science Research Centers[9] that will provide just such capabilities. Four of these centers are listed here, and one is mentioned later when we discuss biological capabilities.

The Center for Nanoscale Materials is focused on fabricating and exploring novel nanoscale materials and, ultimately, employing unique synthesis and characterization methods to control and tailor nanoscale phenomena.

The Center for Functional Nanomaterials provides state-of-the-art capabilities for the fabrication and study of nanoscale materials, with an emphasis on atomic-level tailoring to achieve desired properties and functions.

The Center for Integrated Nanotechnologies features low vibration for sensitive characterization, chemical/biological synthesis labs, and clean room for device integration.

The Center for Nanophase Materials Sciences is a collaborative nanoscience user research facility for the synthesis, characterization, theory/modeling/simulation, and design of nanoscale materials.

Other agencies and even some universities support key nanofabrication facilities. The National Science Foundation funds several nanofabrication facilities, such as at Cornell University, that are available to external users, and which are part of a larger National Nanotechnology Infrastructure Network[10] (NNIN). The Cornell Nanofabrication Facility[11] provides fabrication, synthesis, characterization, and integration capabili-

---

[9]*http://www.science.doe.gov/Sub/Newsroom/News_Releases/DOE-SC/2006/nano/index.htm.*
[10]*http://www.nnin.org.*
[11]*http://www.cnf.cornell.edu.*

ties to build structures, devices, and systems from atomic to complex large scales. Carnegie Mellon University independently operates its own user facility that serves the broader community. The Nanofabrication Facility at Carnegie Mellon[12] provides facilities for data storage thin film and device development and includes extensive clean-room space.

5.2.c Genetics, Proteomics, and Biological Engineering

Biological engineering capabilities are increasingly important to chemical engineers. A few examples of new centers providing state-of-the-art facilities and approaches are given below—starting with one of the Department of Energy nanoscale science research centers.

The Molecular Foundry[13] provides instruments and techniques for users pursuing integration of biological components into functional nanoscale materials.

The Institute for Systems Biology[14] takes a multidisciplinary approach to addressing systems biology that includes integration of research in many sciences including biology, chemistry, physics, computation, mathematics, and medicine.

The Broad Institute[15] brings together research groups with a shared commitment to important biomedical challenges, along a set of key "platforms": biological samples, genome sequencing, genetic analysis, chemical biology, proteomics, and RNAi.

The Synthetic Biology Engineering Research Center (SynBERC)[16] focuses on synthetic biology, fabricating new biological components and assembling them into integrated, miniature devices and systems.

5.2.d Fossil Fuel Utilization Facilities (Combustion Centers)

Chemical engineers have long required capabilities for understanding combustion and fossil fuel utilization. A few examples of centers providing state-of-the-art facilities and approaches are given below.

---

[12]*http://www.nanofab.ece.cmu.edu.*
[13]*http://foundry.lbl.gov/.*
[14]*http://www.systemsbiology.org/.*
[15]*http://www.broad.harvard.edu/.*
[16]*http://www.synberc.org.*

The Combustion Research Facility (CRF) at the Sandia National Laboratories in Livermore[17] is a Department of Energy Office of Science user facility, conducting basic and applied research that has pioneered the use of laser diagnostics for in situ measurements in a wide range of furnace and engine applications.

The Building and Fire Research Laboratory at NIST [18] has unique facilities and programs for addressing the needs of the building and fire safety communities and provides science standards developments, metrology for standards, and responses to major fires using its full-scale fire laboratory.

The International Flame Research Foundation at Livorno, Italy,[19] is a cooperative international organization focusing on applied combustion research and serves industry and academia, with 10 national committees, including the American Flame Research Committee, and excellent facilities at the ENEL plant outside of Pisa.

5.2.e Cyberinfrastructure (Supercomputing)

According to the National Science Foundation, cyberinfrastructure refers to the distributed computer, information, and communication technologies combined with the personnel and integrating components that provide a long-term platform to empower the modern scientific research endeavor.[20] Two examples of engineering cyberinfrastructure capabilities include:

The Collaborative Large-scale Engineering Analysis Network for Environmental Research (CLEANER)[21] addresses large-scale human-stressed aquatic systems through collaborative modeling and knowledge networks.

The Network for Computational Nanotechnology[22] connects theory, experiment, and computation in a way that makes a difference to the future of nanotechnology.

---

[17]*http//www.ca.sandia.gov/CRF.*
[18]*http//www.bfrl.nist.gov.*
[19]*http//www.ifrf.net.*
[20]See extensive list of links on cyber-infrastructure at *http://www.nsf.gov/crssprgm/ci-team/ #ecI.*
[21]*http://cleaner.ncsa.uiuc.edu/home/.*
[22]*http://www.ncn.purdue.edu/.*

5.3 HUMAN RESOURCES

Human resources are an essential component for leadership in chemical engineering. Below we discuss trends and several key characteristics of science and engineering human resources in the world overall, and then drill down into some important features of the U.S. supply of chemical engineers.

5.3.a Strong Competition for International Science and Engineering Human Resources

At the international level, the United States ranks lower than most industrialized nations in terms of the quantity of natural sciences and engineering degrees awarded per number of 24-year-olds in the general population (Figure 5.1). Many more overall science and engineering (S&E) degree holders are being produced abroad than in the United States. However, over the years, the United States has been successful at attracting foreign-born scientists and engineers (Figure 5.2).

5.3.b Steady Supply of Chemical Engineers in the United States

It is difficult to find numbers for chemical engineering human resources at the international level. The best we can do is look at the trends in U.S. chemical engineering graduate degrees to get some indication of the current health of the discipline and where things are headed.

Over the period 1983-2004 (shown in Figure 5.3), there has been an overall steady supply of graduate students enrolling in chemical engineering. However, if we look more carefully at the residence status of graduate students, there has been a significant decrease in the number of U.S. citizens/permanent residents enrolling in chemical engineering graduate programs. As it turns out, the decrease has been made up by enrollment of temporary residents.

A better indicator of current trends, however, is to look at first-time full-time graduate enrollments, because overall graduate student enrollments include individuals who began school up to 5 or 6 years ago. We see that since the mid 1980s, first-time full-time graduate student enrollments in the United States (Figure 5.4) have fluctuated, but have overall remained constant. At the same time, recently reported numbers from National Science Foundation show a nearly 13% decrease in enrollment of first-time full-time chemical engineering graduate students.

Since we are most interested in competitiveness of chemical engineering research, it is critical to look at the supply of PhDs. We see in Figure 5.5 that between the late 1970s and early 1990s, the number of earned chemi-

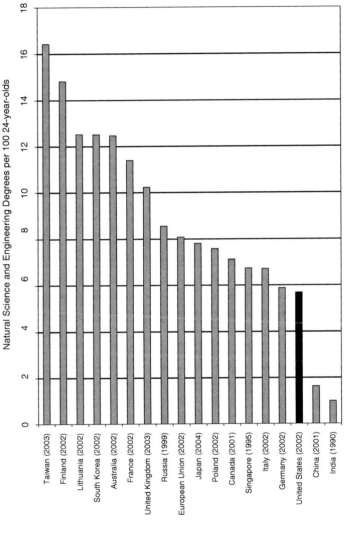

FIGURE 5.1 Natural science and engineering (NS&E) degrees per 100 24-year-olds by country/economy, most recent year.
SOURCE: Science and Engineering Indicators 2006 based on data from Organisation for Economic Co-operation and Development, Center for Education Research and Innovation, Education database, *http://www1.oecd.org/scripts/cde/members/edu_uoeauthenticate.asp*; United Nations Educational, Scientific, and Cultural Organization (UNESCO), Institute for Statistics database, *http://www.unesco.org/statistics/*; and national sources.

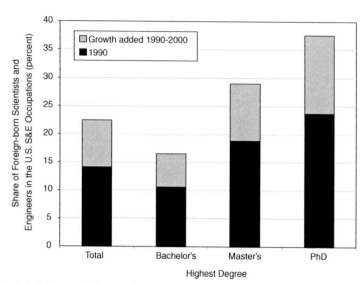

FIGURE 5.2 Share of foreign-born scientists and engineers in U.S. S&E occupations, by degree level, 1990 and 2000.
NOTE: Data exclude postsecondary teachers because of census occupation coding.
SOURCE: Science and Engineering Indicators 2006 based on data from U.S. Census Bureau, 5-Percent Public-Use Microdata Sample, *http://www.census.gov/main/www/pums.html.*

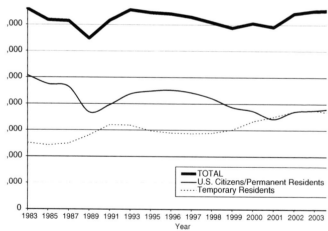

FIGURE 5.3 Total graduate enrollment in chemical engineering and enrollments based on residency status: U.S. citizens/permanent residents versus temporary residents, 1993-2004.
SOURCE: Science and Engineering Indicators 2006, Appendix Table 2-15; and National Science Foundation, Division of Science Resources Statistics, *Graduate Students and Postdoctorates in Science and Engineering: Fall 2004,* NSF 06-325, Project Officer, Julia D. Oliver (Arlington, VA 2006).

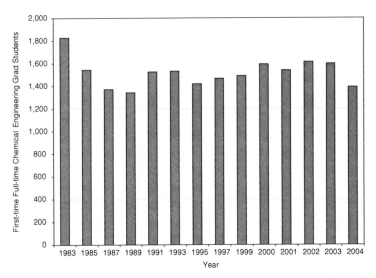

FIGURE 5.4 First-time full-time graduate student enrollments for chemical engineering, 1983-2004.

SOURCE: S&E Indicators 2006, Appendix Table 2-13 and National Science Foundation, Division of Science Resources Statistics, *Graduate Students and Postdoctorates in Science and Engineering: Fall 2004*, NSF 06-325, Project Officer, Julia D. Oliver (Arlington, VA 2006).

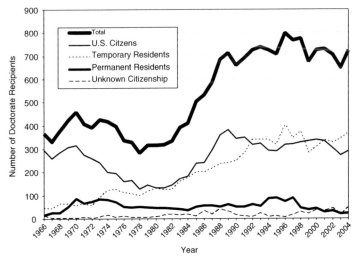

FIGURE 5.5 Earned doctoral degrees in chemical engineering from U.S. institutions as a function of residency status, 1966-2004.

SOURCE: NSF/SRS, Survey of Earned Doctorates, Integrated Science and Engineering Resources Data System (WebCASPAR), *http://webcaspar.nsf.gov* (accessed September 5, 2006).

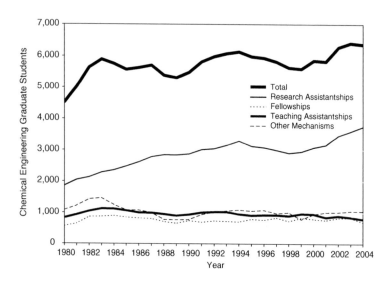

FIGURE 5.6 Chemical engineering graduate students by mechanism of support, 1980-2004.
SOURCE: NSF/SRS, Survey of Earned Doctorates, Integrated Science and Engineering Resources Data System (WebCASPAR), *http://webcaspar.nsf.gov* (accessed September 5, 2006).

cal engineering PhDs in the United States grew quite rapidly and more than doubled, largely due to increased numbers of doctorates awarded to temporary residents. Over the past 10 years (1994-2004), the number of earned chemical engineering doctorates awarded each year has fluctuated slightly, but overall has remained fairly level at around 700 doctorates awarded per year. In comparison, for the 213 non-U.S. chemical engineering departments who provided data to the University of Texas, Austin, Chemical Engineering Faculty Directory for the years 2003-04 or 2004-05, there were 1923 PhD degrees awarded.[23]

Graduate students in chemical engineering have been supported adequately over the past 20 years. During this time period, graduate research assistantships have increased significantly. Research assistantships accounted for more than 50% of graduate student support in 2004 (see Figure 5.6).

Approximately half of all chemical engineering graduate students are supported by research assistantships. A large number of these assistantships are funded by federal agencies such as the National Science Foundation (Figure 5.7).

---

[23] *http://www.che.utexas.edu/che-faculty/index.html.*

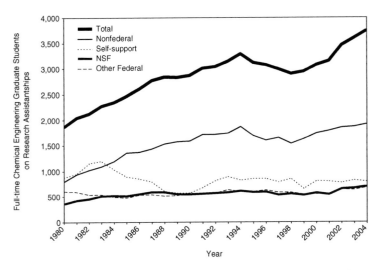

FIGURE 5.7 Full-time graduate students in chemical engineering on research assistantships, by funding source, 1980-2004. NOTE: NSF = National Science Foundation. SOURCE: NSF/SRS, Survey of Earned Doctorates, Integrated Science and Engineering Resources Data System (WebCASPAR), http://webcaspar.nsf.gov (accessed September 5, 2006).

5.3.c Job Prospects and Salaries for U.S. Chemical Engineers Are Still Favorable

The number of employed chemical engineering degree holders has steadily increased (Figure 5.8). The percentage increase from 1999 to 2003 was 8% overall, 4% for bachelor's, 19% for master's, and 17% for Ph.D.'s.

Figure 5.9 shows that there was also an increase in the number of employed chemical engineering degree holders across all employment sectors. However, the fraction of individuals employed by the business sector fell from 88% to 84%, while the percentage employed by the education and government sectors increased respectively from 4% to 7%, and 8% to 9%.

However, there has been a change in where chemical engineers (not necessarily chemical engineering degree holders) are employed. Figure 5.10 below shows the decline of chemical engineers being employed in the chemical industry and the concomitant growth in the electronics industry.[24]

---

[24]E. L. Cussler and J. Wei, Chemical product engineering, *AIChE Journal* 49(5):1072-1075 (2003).

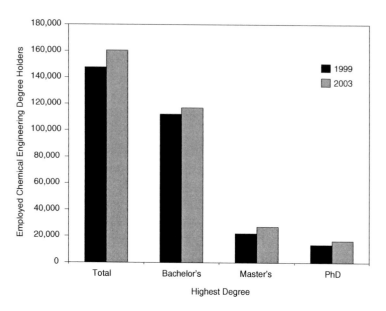

FIGURE 5.8 Comparison of employed chemical engineering degree holders, 1999 and 2003.
SOURCE: 2004 and 2006 S&E Indicators.

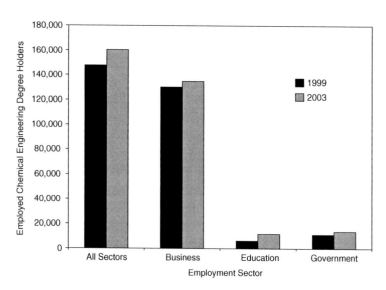

FIGURE 5.9 Comparison of employed chemical engineering degree holders across different sectors, 1999 and 2003.
SOURCE: 2004 and 2006 S&E Indicators.

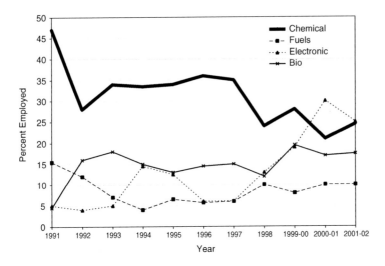

FIGURE 5.10 Trends in employment of doctoral level chemical engineers in various industries, 1991-2001.
SOURCE: E. L. Cussler and J. Wei, "Chemical Product Engineering," *AIChE Journal*, 49, no. 5 (2003).

According to the Bureau of Labor Statistics' 2006-2007 *Occupational Outlook Handbook*[25] the job prospects for those employed as chemical engineers (not necessarily chemical engineering degree holders) over the next 5-10 years looks quite good. While engineers are employed in every major industry, as expected, the chemical industry employs the largest percentage of chemical engineers (27.8%), followed by architectural, engineering, and related services industries (16.3%). Chemical engineers are expected to have employment growth about as fast as the average (9% to 17%) for all occupations through 2014. They state that although overall employment in the U.S. chemical manufacturing industry is expected to decline, chemical companies will continue to carry out R&D on new chemicals and more efficient processes to increase output of existing chemicals. At the same time, the handbook says that among manufacturing industries, pharmaceuticals may provide the best opportunities for jobseekers and that most employment growth for chemical engineers will be in service industries, such as scientific research and development services, particularly in energy and the developing fields of biotechnology and nanotechnology.

---

[25]Bureau of Labor Statistics, Occupational Outlook Handbook, 2006-2007 edition. In *Engineers*, U.S. Department of Labor, Washington, D.C., 2006.

TABLE 5.1 Average Starting Salary Offers for Engineers

Curriculum	Bachelor's	Master's	PhD
Aerospace/aeronautical/astronautical	$50,993	$62,930	$72,529
Agricultural	46,172	53,022	—
Bioengineering and biomedical	48,503	59,667	—
Chemical	53,813	57,260	79,591
Civil	43,679	48,050	59,625
Computer	52,464	60,354	69,625
Electrical/electronics and communications	51,888	64,416	80,206
Environmental/environmental health	47,384	—	—
Industrial/manufacturing	49,567	56,561	85,000
Materials	50,982	—	—
Mechanical	50,236	59,880	68,299
Mining & mineral	48,643	—	—
Nuclear	51,182	58,814	—
Petroleum	61,516	58,000	—

SOURCE: Bureau of Labor Statistic 2006-2007 Occupational Outlook Handbook, based on 2005 survey by the National Association of Colleges and Employers.

The handbook also discusses expected earnings for chemical engineers. While earnings for engineers vary significantly by specialty, industry, and education, engineers as a group earn some of the highest average starting salaries among those holding bachelor's degrees. Table 5.1 shows the current average starting salary offers for engineers, with chemical engineers ranking among the most highly paid degree holders.

Data from the American Chemical Society 2004 Survey on Starting Salaries of Chemists and Chemical Engineers (Figure 5.11) shows that starting salaries for chemical engineers have steadily increased since 1975. However, this increase (4.74% average annually) has just barely kept pace with inflation.[26]

Earnings for more experienced chemical engineers (with PhDs) as measured by median annual salary since degree (Figure 5.12) has grown a bit more than starting salaries (3.7% annually), but has also barely kept up with inflation.[27]

---

[26]Consumer Price Index, average annual increase for 1975-2004 is 4.42% (Bureau of Labor Statistics Inflation Calculator data.bles.gov/cgi-bin/cpicalc.pl accessed 9-8-06, a dollar in 1975 is equivalent to $3.51 in 2004).

[27]The average annual increase in the consumer price index for 1993-2003 was 2.42% (according to the Bureau of Labor Statistics inflation calculator, $1.00 in 1993 is equivalent to $1.27 in 2003).

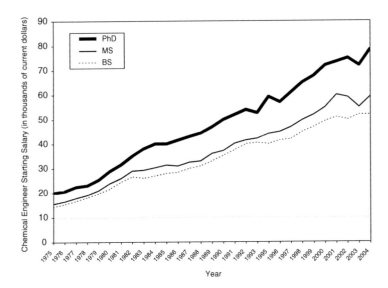

FIGURE 5.11 Inexperienced chemical engineer median starting salaries by degree held.
SOURCE: ACS 2004 Survey on Starting Salaries of Chemists and Chemical Engineers.

5.4 R&D FUNDING

Here we look at trends in international levels of S&E funding and specific R&D funding for chemical engineering in the United States. As discussed earlier, the U.S. innovation system benefits greatly from the variety and well as the consistency of funding sources.

5.4.a Steady Funding for S&E in the United States

The United States has spent more on science and engineering R&D over the time period of 1981-2002 than any other Organisation for Economic Co-operation and Development (OECD) country (Figures 5.13 and 5.14). In 2003, the United States spent more than $250 billion (constant 2000 $US) on total R&D. The United States accounted for more than 40% of the yearly international expenditures for S&E. Between 1981 and 2001, the U.S. contribution declined from 45% to 43%, and the G7 contribution declined from 91% to 84%.

Because of the differences in the size and economies of different nations, it is useful to normalize R&D expenditures based on gross domestic prod-

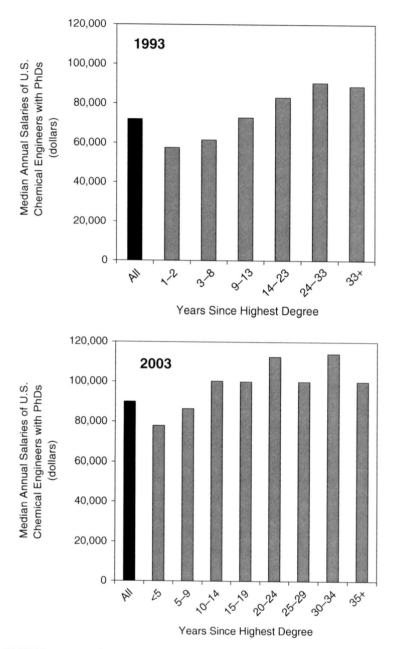

FIGURE 5.12 Median annual salaries for chemical engineers with PhDs by years since highest degree received, 1993 and 2003.
SOURCE: National Science Foundation/SRS, 1993 & 2003 Survey of Doctorate Recipients.

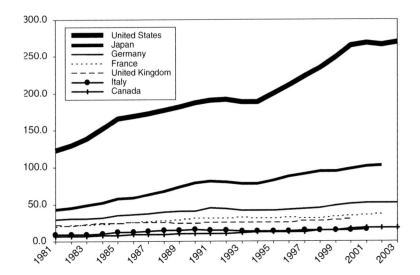

FIGURE 5.13 International R&D expenditures for G7 countries, 1981-2003 in billions of constant 2000 U.S. dollars.
SOURCE: Appendix Table 4-42, Science and Engineering Indicators 2006.

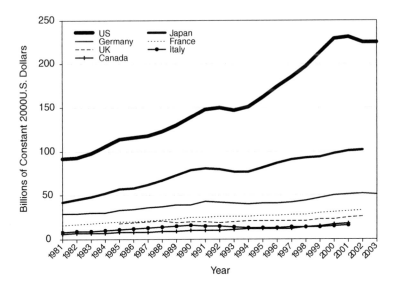

FIGURE 5.14 International nondefense R&D expenditures for select countries, 1981-2003.
SOURCE: Appendix Table 4-43, NSF S&E Indicators 2006.

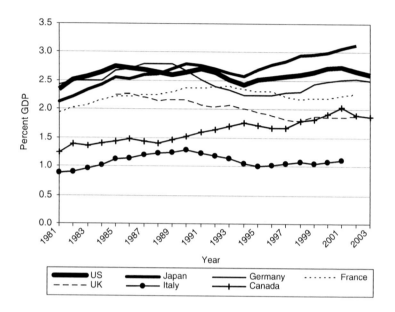

FIGURE 5.15 International R&D as a percentage of gross domestic product by selected country, 1981-2003.
SOURCE: Appendix Table 4-42, S&E Indicators 2006.

uct (GDP). As seen in Figures 5.15 and 5.16, the United States is among the leaders in gross domestic expenditures on R&D, ranking fifth among OECD countries in terms of reported R&D/GDP ratios.[28] However, Israel (not an OECD country), devoting 4.9% of its GDP to R&D, led all countries, followed by Sweden (4.3%), Finland (3.5%), Japan (3.1%), and Iceland (3.1%).[29] Although China reported R&D expenditures similar to Germany in 2000, on a per capita basis, Germany's R&D was over 16 times that of China.

As in most of the developed nations (Figures 5.17 and 5.18), the industrial sector in the United States spends the most on and performs most of the R&D. Industry funds about 60% of the R&D, and the federal government funds about 30%. However, industry conducts nearly 70% of R&D,

---

[28] As noted by the National Science Foundation, "Growth in the R&D/GDP ratio does not necessarily imply increased R&D expenditures. For example, the rise in R&D/GDP from 1978 to 1985 was due as much to a slowdown in GDP growth as it was to increased spending on R&D activities."

[29] See NSF S&E Indicators 2006, Table 4-13.

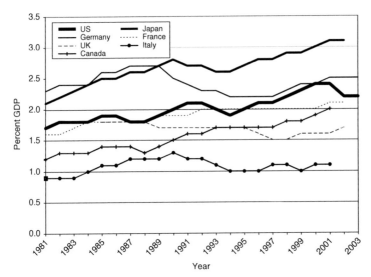

FIGURE 5.16 International nondefense R&D as a percentage of GDP, by selected country, 1981-2003.
SOURCE: Appendix Table 4-43, S&E Indicators 2006.

while the rest is split among higher education (15%), government (10%), and private/nonprofit (5%).

Compared with other countries that support a substantial level of academic R&D (at least $1 billion purchasing power parity in 1999), the United States devotes a smaller proportion (15%) of its R&D to engineering and social sciences. However, in terms of the actual expenditures for engineering, the United States leads the other industrialized nations (Figure 5.19).

5.4.b Steady U.S. Funding for Chemical Engineering R&D

In 2004, nearly $500 million was spent on chemical engineering R&D at academic institutions (Figure 5.20). Of this, about 54% was from federal sources.

In terms of constant 2000 dollars, the U.S. federal obligations for total research in chemical engineering declined from a high of about $ 350 million in 1992 to about $ 200 million in 2002 (Figure 5.21). More recently, the numbers have increased to about $300 million, due to a large increase

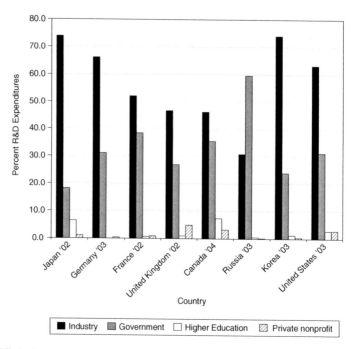

FIGURE 5.17 International R&D expenditures for selected countries, percent distribution by source of funds.
SOURCE: Appendix Table 4-44, S&E Indicators 2006.

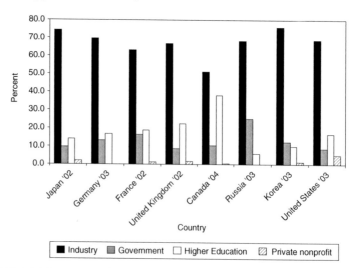

FIGURE 5.18 International R&D expenditures for selected countries, percent distribution by performing sector.
SOURCE: Appendix Table 4-44, S&E Indicators 2006.

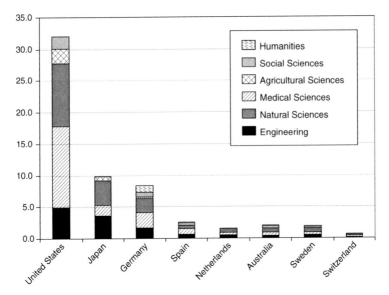

FIGURE 5.19 Share of academic R&D expenditures, by country and S&E field: Selected years, 2000-2002.
SOURCE: Table 4-14, Science and Engineering Indicators 2006.

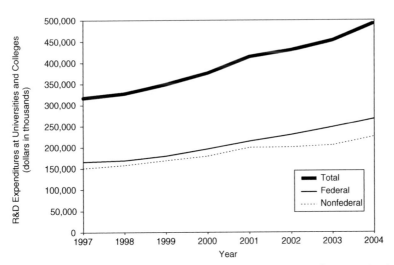

FIGURE 5.20 Federal and nonfederal R&D expenditures at academic institutions for chemical engineering.
SOURCE: National Science Foundation/Division of Science Resources Statistics, Survey of Research and Development Expenditures at Universities and Colleges, FY 2004.

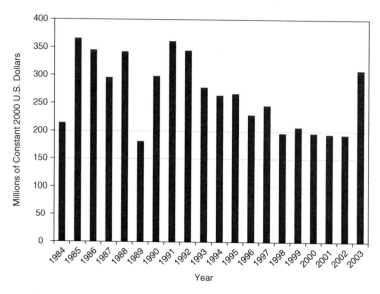

FIGURE 5.21 Federal obligations for total research in chemical engineering. SOURCE: Appendix Table 4-32, Science and Engineering Indicators 2006.

in funding from the Department of Energy (shown later in Figure 5.23). Federal obligations for chemical engineering over the time period (1984-2003) ranges from a low of 0.4% in 2000 and 2001 to 1.6% of the total U.S. R&D budget in 1985.

The federal funding for chemical engineering research is comparable in spending with two of the other "big four" engineering fields of civil and mechanical engineering—electrical engineering has traditionally been better funded than the other three (Figure 5.22).

5.4.c A Changing Landscape for Chemical Engineering R&D Funding

The different federal agency contributions to the total funding for chemical engineering research are shown in Figure 5.23. The Department of Energy has made the largest overall contribution to chemical engineering research over the 20 years shown. DOE funding was at a maximum of about $142 million in 1991, dropped to $92 million in 2002, and jumped to $198 million in 2003.

Below is a comparison of Department of Energy Basic Energy Sciences funding for core research areas in chemistry, geosciences, and biosciences (Figure 5.24) and materials (Figure 5.25) for fiscal year 2001 and fiscal year

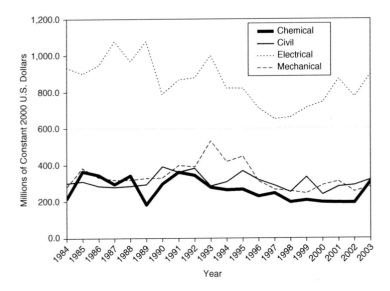

FIGURE 5.22 Federal obligations for total research, by engineering field—"The Big Four": fiscal year 1984-2003.
SOURCE: Appendix Table 4-32, Science and Engineering Indicators 2006 Academic R&D Expenditures.

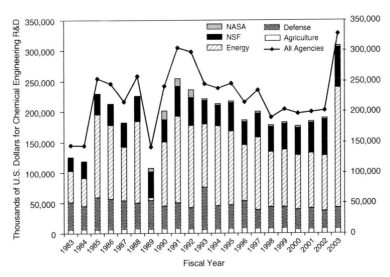

FIGURE 5.23 Federal obligations for total chemical engineering research, by select agency, fiscal years 1984-2003.
SOURCE: National Science Foundation, *Federal Funds for R&D*.

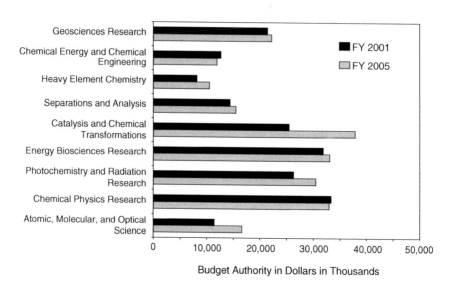

FIGURE 5.24 Department of Energy Basic Energy Sciences funding for Chemical, Geological, and Biological Core Research Activities.
SOURCE: *http://www.er.doe.gov/bes/brochures/CRA.html.*

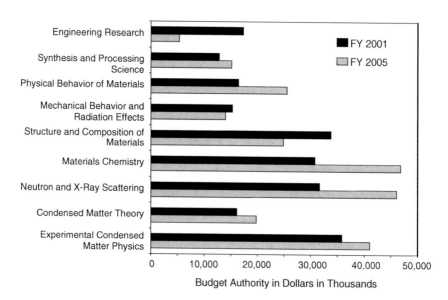

FIGURE 5.25 Department of Energy Basic Energy Sciences funding for Material Science and Engineering Core Research Activities.
SOURCE: *http://www.er.doe.gov/bes/brochures/CRA.html.*

2005. There was a large increase ($10 million) for catalysis and chemical transformations, as well as modest increases for atomic, molecular, and optical science and photochemistry and radiation research.

Federal academic research obligations for chemical engineering are less balanced among agencies than 10 years ago (Figure 5.26). The National Science Foundation now accounts for 66% of the federal academic research obligations for chemical engineering. Ten years ago a larger proportion of R&D funding for chemical engineering came from the Department of Energy.

The National Institutes of Health does not appear in figure 5-26 as one of the major funding agencies for academic chemical engineering research. However, the five year doubling of the NIH budget between 1998 and 2003 has significantly increases NIH's contribution to chemical engineering departments (Figure 5-27).

Figure 5.28 shows the breakdown of funding for the divisions of the National Science Foundation Engineering Directorate. The Chemical Transport Systems (CTS) Division mainly supports chemical engineering research at academic institutions.

Recently, CTS was joined with the Bioengineering and Environmental Systems (BES) Division to create the Chemical, Bioengineering, Environment, & Transport (CBET) Division. Table 5.2 shows the overall research proposal funding rate for CBET. While, the number of awards has remained

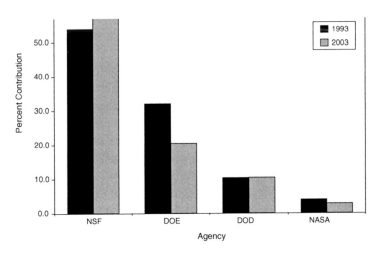

FIGURE 5.26 Federal academic research obligations for chemical engineering provided by major agencies.
SOURCE: Appendix Table 5.09, S&E Indicators 2006 and Appendix Table 5.11, S&E Indicators 1996.

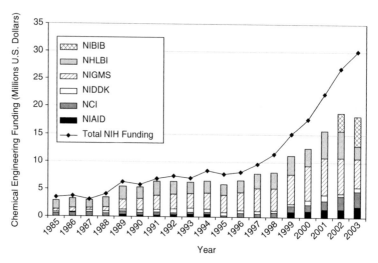

FIGURE 5-27 NIH support for chemical engineering department programs by institute, 1985-2003.
NOTE: NIBIB = National Institute of Biomedical Imaging and Bioengineering; NHLBI = National Heart, Lung, Blood Institute; NIAID = National Institute of Allergy and Infectious Diseases; NCI = National Cancer Institute; NIDDKD = National Institute of Diabetes and Digestive and Kidney Diseases; NIGMS = National Institute of General Medical Sciences.
SOURCE: National Institute of General Medical Sciences Office of Program Analysis and Evaluation compilation of biochemistry, chemistry, and chemical engineering department support based on data from the NIH IMPAC system.

fairly stable and the median annual size of awards has increased between 1997 and 2005, the funding rate for awards has substantially decreased. (For similar data for CBET funding areas see table in Appendix 5A at end of this chapter.)

5.5 PROJECTION OF LEADERSHIP DETERMINANTS

In this section, we attempt a projection of the leadership determinants, which underpin the likelihood of predictions made in earlier Chapters 3 and 4.

5.5.a Recruitment of Talented Researchers

U.S. institutions continue to attract and retain the world's best scientists and engineers because of the presence of other outstanding researchers

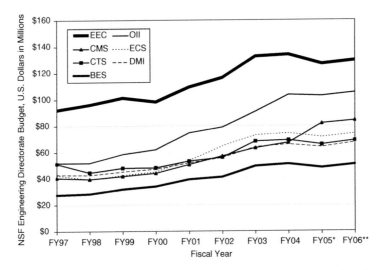

FIGURE 5.28 National Science Foundation Engineering Directorate funding for divisions in millions of U.S. dollars: Bioengineering and Environmental Systems (BES), Chemical and Transport Systems (CTS), Civil and Mechanical Systems (CMS), Design and Manufacturing Innovation (DMI), Electrical and Communications Systems (ECS), Engineering Education and Centers (EEC), Office of Industrial Innovation (OII).
NOTE: *FY05 planned budget; **FY06 proposed budget.
SOURCE: NSF FY06 Budget request, available at *http://www.nsf.gov/about/budget* (accessed October 5, 2006).

TABLE 5.2 Research Proposal Funding Rate for National Science Foundation Chemical, Bioengineering, Environment & Transport (CBET) Division from Fiscal Year 1997 to 2005.

Fiscal Year	Number of Proposals	Number of Awards	Funding Rate (%)	Median Annual Size ($)
2005	2,712	353	13	94,124
2004	2,084	421	20	87,188
2003	1,962	397	20	86,816
2002	1,449	403	28	79,818
2001	1,449	374	26	79,994
2000	1,459	410	28	75,000
1999	1,122	364	32	69,035
1998	1,267	379	30	64,400
1997	1,363	413	30	57,523

SOURCE: National Science Foundation Budget Internet Information System *http://dellweb.bfa.nsf.gov/* (assessed October 6, 2006).

with whom these individuals work, a superior economy, and outstanding research facilities. Evidence of this is the level to which foreign doctorate recipients plan to remain in the United States to work after graduation (Table 5.3). However, with changes in visa policies (such as the drop in student visas issued after 9/11 shown in Figure 5.29) and global leveling in research capability, the United States may be losing ground.

The data presented so far raise many issues that affect the future ability of chemical engineering programs to attract high-quality graduate students, and include

- Recruiting students from both within the United States and abroad. The decreasing numbers of U.S. citizens or permanent residents attending Ph.D. programs is worrisome.
- Improving and strengthening academic programs so they can still remain poles of attraction for young people with intellectual curiosity.
- Retaining an open and active research environment, which has been one of the most attractive features, especially for non-U.S. prospective Ph.D. students.
- Ensuring adequate financial support for U.S. students pursuing graduate education.
- Maintaining a strong job market for chemical engineering graduates (especially PhDs) with improved incentives and more attractive career paths.
- Increasing diversity in academia, government, and industry chemical engineering leadership.

5.5.b R&D Funding

Whereas U.S. industry and government are shifting funds toward shorter-term research, many other countries, notably Japan, are increasing long-term and basic research funding. Many U.S. companies have eliminated or significantly reduced in size corporate or central research

TABLE 5.3 Percentage of Foreign Doctorate Recipients Reporting Plans to Stay in the United States After Graduation, 1995-2003

	1994	1995	1996	1997	1998	1999	2000	2001	2002	2003
Definite Plans to Stay	34	35	42	44	46	49	49	54	52	48
Plans to Stay	62	65	67	68	67	70	71	74	73	71

SOURCE: Special Tabulation of Data from the Survey of Doctorate Recipients, prepared by National Opinion Research Center.

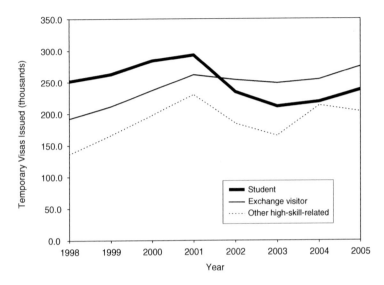

FIGURE 5.29 Student, exchange visitor, and other high-skill-related temporary visas issued, 1998-2005.
SOURCE: NSF 2006 Science & Engineering Indicators.

laboratories in order to more closely align research and development with shorter-term business opportunities.

Chemical engineering research in universities has been sponsored mainly by the federal government. The National Science Foundation and the Department of Energy have provided the most support for a range of fundamental chemical engineering research. In particular, the National Science Foundation now dominates support for chemical engineering with 66% of academic research in the field.

The overall federal research and development funding strategy for chemical engineering research is currently unbalanced. As a result, important developments in key subareas could lag behind in world competition. As was discussed in Chapter 4, several core areas of chemical engineering research are at serious risk. The dynamic range of the discipline, which has been a principal strength for more than 50 years, is seriously threatened by reductions in support of core research areas. This is illustrated by the funding data shown in the table in Appendix 5A. For the areas in the table, the funding rates have dropped to less than half their peak levels over the last 4-5 years. Biophotonics is the only area that has kept a constant funding rate. The overall drop in rates occurred despite a large number of proposals in 2005 (over 600) in biotechnology and biomedical engineering that are far more than the submissions 4-5 years ago.

Although some academic researchers have turned to industry for financial support, in many cases, industry-funded research is of shorter duration and, compared with federal grants, has a specific, short-term focus. Some research projects are conducted under contract terms that capture intellectual properties, protect confidentiality, restrict publication, and require detailed planning and reporting of progress. These conditions rarely attract top graduate talent to the research effort.

In Chapter 4 we discussed areas in which industrial research collaborations can be most valuable, where special equipment not generally found in universities is required to achieve process control and to evaluate sequencing protocols and scaling parameters.

5.5.c Infrastructure

The quality of the basic research infrastructure and the development of new technology from research strongly influence the long-term health of chemical engineering research. The position of the U.S. research enterprise will be determined by the elevation or decline of this infrastructure, which, in this context, is defined broadly to include tangible (facilities) and intangible (supporting policies and services) elements. Several trends for the elements of this infrastructure have been identified:

The university structure in which the chemical engineering organization resides strongly influences the fortunes of the discipline. The high quality of academic leadership in chemical engineering and the excellence of the engineering research enterprise have placed the discipline in a position of strength at most of the top research universities in the United States. The prominence of chemical engineering in nonacademic institutions (industry and government agencies) is also well established here and abroad.

Major centers and facilities provide key infrastructure and capabilities for conducting research and have provided the foundation for U.S. leadership. Key capabilities for chemical engineering research include materials synthesis and characterization, materials micro- and nanofabrication, genetics and proteomics, fossil fuel utilization, and cyberinfrastructure. U.S. facilities have instrumentation that is on par with the best in the world. However, rapid advances in design and capabilities of instrumentation can create obsolescence in 5-8 years.

Forward-looking intellectual property policies, administrative support, and access to patent expertise are improving for U.S. academic researchers in chemical engineering. These policies are generally more flexible and advanced here than they are abroad. The anticipated continuing liberalization of rules that permit academic researchers to commercialize their inventions is a positive step toward decreasing the time from invention to market. Another positive step is the growing assistance from the universities in finding industrial commercialization partners.

Federal laboratories and the national laboratories of the Department of Energy are critical in providing unique facilities for research; they have instrumentation no single university could afford to put in place. An important complement is the availability of world-class scientists who engage in long-term fundamental research, provide assistance through research collaborations with the user community, and provide advanced instrumentation design and methods. Large central facilities, such as neutron and synchrotron sources, electron microscopy centers, and analytical facilities, many of them at Department of Energy laboratories, must be continuously upgraded and maintained.

Although the United States has enjoyed a research and funding environment that allows for the installation and operation of a diverse range of facilities to support leading-edge research in chemical engineering, this position is not assured forever.

5.5.d Cooperative Government-Industry-Academia Research

Maintaining a competitive advantage in chemical engineering depends on strong collaborations between government, industry, and academia. As industrial research focuses more and more on short-term (2-3 year) targeted advances and product impact, execution of longer term (5-10 year) basic and innovative exploratory research at universities and national laboratories will require even closer interactions. Collaborative research is accomplished in several foreign countries by individuals with joint academic-commercial appointments and through publicly supported research institutes linked to universities (similar to many U.S. national laboratories) that serve industry's need for longer-term research.

One challenge is also a major opportunity for a government-university-industry initiative: There is a 15-year cycle time in many cases from demonstrating the scientific feasibility of a new idea to its commercial implementation. There is a need for continuity of support and a general recognition of the time it takes to go from observation to hypothesis to experimentation to discovery to implementation. A reduction in this schedule could be realized through more extensive integration of modeling and simulation of the processes with evaluation of fabrication concepts and designs, processing yields, performance, and reliability. There are clearly defined, mutually supportive roles for academia, government, and industry where they can work together. For example, the Department of Energy advanced supercomputer initiative is an effort to develop new computer methods for the simulation of nuclear weapons. Analogous models of cooperative government-industry-academia research may be needed to enhance the transfer of results from fundamental research to viable engineering solutions in the new and evolving areas.

5.5.e Government Policy and Regulations

Government policy and regulations have a direct impact on the choice of directions and intensity of chemical engineering research by industry and academia. They affect cost of raw materials (e.g., natural gas), influence research undertakings (e.g., biorefineries, fuels from cellulose), determine the scope of new technologies (e.g., processes and materials for tighter control of air and water effluents), and encourage or discourage the introduction of new materials in the market (e.g., regulations governing the approval of new biomedical devices and the litigation-based culture in the United States).

Most of our analysis in forecasting the future position of U.S. research in chemical engineering has been predicated on rather "neutral" new regulations. However, the Panel believes that this is a question of significant uncertainty and with enormous impact on the directions and position of future chemical engineering research.

5.6 SUMMARY AND CONCLUSIONS

Historical research leadership in chemical engineering in the United States is the result of many key factors, which have been outlined in this chapter.

Over the years, the United States has been a leader in *innovation* as a result of a strong U.S. industrial sector, a variety of funding opportunities (industry, federal government, state initiatives, universities, and private foundations), cross-sector collaborations and partnerships, and strong professional societies. While U.S. chemical companies will retain a very strong presence in the global market, the corresponding size of their operations from the U.S. market will grow at a rather low rate. In time, it may have an impact on the number and type of employment opportunities offered to U.S. chemical engineering researchers and the cultivation of research initiatives in collaboration with U.S. universities.

Major centers and facilities provide key infrastructure and capabilities for conducting research, and have provided the foundation for U.S. leadership. Key capabilities for chemical engineering research include materials synthesis and characterization, materials micro- and nanofabrication, genetics and proteomics, clean and efficient fossil fuel utilization, renewable energy sources, and cyberinfrastructure.

In the past, the United States was well endowed with *human resources* in science and engineering. There has been an overall steady supply of chemical engineers in the United States, and job prospects and salaries for U.S. chemical engineers are still favorable. However, with changes in U.S. citizenry interests and international capabilities, there is increasingly strong competition for international science and engineering human resources. Other professions offer higher monetary compensation and attractive career paths, which help them draw talented young people away from science and engineering education or away from science- or engineering-oriented employment positions. Most major companies are building new R&D centers outside the United States, such as in China and India. For example, DuPont recently announced plans to invest over $22.5 million to construct its first research and development center in Hyderabad, India, which is expected to accommodate more than 300 scientists and other employees.[30] Additional examples include GE (India and China), Dow Chemical (India and China), and Rohm and Haas (China). Citizens of those countries are increasingly gaining access to world-class facilities to work in, which will increasingly be competitive with those in the United States.

Research funding for S&E overall and chemical engineering in particular has been steady over all the years. However, the landscape for chemical engineering has changed significantly, and a reassessment of funding policy directions may be needed in view of this report's findings.

---

[30]See *http://www2.dupont.com/Media_Center/en_US/daily_news/february/article20070202. html*, last accessed March 5, 2007.

APPENDIX 5A

Research Proposal Funding Rate for National Science Foundation Chemical, Bioengineering, Environment & Transport (CBET) Division Research Areas from fiscal year 1997 to 2005. SOURCE: NSF Budget Internet Information System, *http://dellweb.bfa.nsf.gov* (accessed October 6, 2006).

CBET Funding Areas	Fiscal Year	Number of Proposals	Number of Awards	Funding Rate	Median Annual Size
BIOCHEMICAL & BIOMASS ENG	2005	34	5	15%	$111,685
	2004	49	8	16%	$90,000
	2003	84	11	13%	$100,000
	2002	57	15	26%	$111,636
	2001	61	21	34%	$79,544
	2000	66	19	29%	$108,400
	1999	75	27	36%	$81,866
	1998	60	20	33%	$69,932
	1997	63	20	32%	$62,500
BIOMEDICAL ENGINEERING	2005	301	32	11%	$100,000
	2004	324	33	10%	$100,500
	2003	218	30	14%	$79,978
	2002	282	37	13%	$76,683
	2001	248	45	18%	$76,198
	2000	265	66	25%	$75,086
	1999	164	45	27%	$65,143
	1998	159	53	33%	$54,593
	1997	158	40	25%	$51,912
BIOPHOTONICS PROGRAM	2005	42	9	21%	$110,000
	2004	27	7	26%	$100,000
	2003	37	9	24%	$98,247
BIOTECHNOLOGY	2005	374	20	5%	$100,000
	2004	206	30	15%	$138,271
	2003	239	29	12%	$109,242
	2002	116	23	20%	$128,642
	2001	107	35	33%	$99,999
	2000	87	26	30%	$101,490
	1999	59	18	31%	$88,327
	1998	45	21	47%	$85,000
	1997	54	21	39%	$63,847

CBET Funding Areas	Fiscal Year	Number of Proposals	Number of Awards	Funding Rate	Median Annual Size
CATALYSIS AND	2005	161	23	14%	$99,881
BIOCATALYSIS	2004	129	27	21%	$81,325
	2003	116	27	23%	$87,185
	2002	73	26	36%	$74,999
	2001	73	19	26%	$84,000
	2000	96	26	27%	$79,501
	1999	61	25	41%	$70,000
	1998	65	30	46%	$83,073
	1997	84	27	32%	$60,650
COMBUSTION AND	2005	153	17	11%	$76,495
PLASMA SYSTEMS	2004	73	15	21%	$52,500
	2003	78	31	40%	$102,398
	2002	53	23	43%	$80,800
	2001	49	22	45%	$81,334
	2000	75	24	32%	$86,820
	1999	56	23	41%	$82,500
	1998	45	16	36%	$65,012
	1997	69	23	33%	$60,000
ENVIRONMENTAL	2005	306	42	14%	$99,998
ENGINEERING	2004	205	47	23%	$80,001
	2003	273	50	18%	$90,749
	2002	163	40	25%	$80,531
	2001	127	32	25%	$81,393
	2000	59	17	29%	$59,750
	1999	50	13	26%	$65,000
	1998	87	17	20%	$70,305
	1997	118	29	25%	$62,258
ENVIRONMENTAL	2005	51	12	24%	$104,942
TECHNOLOGY	2004	70	12	17%	$81,386
	2003	78	4	5%	$110,689
	2002	54	17	31%	$95,196
	2001	150	26	17%	$76,263
	2000	147	30	20%	$66,845
	1999	133	20	15%	$60,360
	1998	97	34	35%	$50,296
	1997	115	45	39%	$49,931

CBET Funding Areas	Fiscal Year	Number of Proposals	Number of Awards	Funding Rate	Median Annual Size
FLUID DYNAMICS & HYDRAULICS	2005	228	25	11%	$73,690
	2004	104	35	34%	$81,200
	2003	72	16	22%	$80,000
	2002	115	33	29%	$75,000
	2001	134	29	22%	$75,000
	2000	96	38	40%	$70,000
	1999	76	24	32%	$58,125
	1998	99	23	23%	$67,083
	1997	150	29	19%	$62,500
INTERFAC TRANS & THERMODYN PRO	2005	177	19	11%	$90,155
	2004	89	30	34%	$80,000
	2003	186	34	18%	$51,991
	2002	106	43	41%	$83,094
	2001	102	24	24%	$81,054
	2000	117	38	32%	$73,041
	1999	75	35	47%	$62,500
	1998	80	42	53%	$57,305
	1997	111	42	38%	$64,500
PARTICULATE &MULTIPHASE PROCES	2005	267	42	16%	$60,000
	2004	159	47	30%	$80,000
	2003	123	44	36%	$77,703
	2002	97	38	39%	$62,126
	2001	73	33	45%	$69,583
	2000	113	37	33%	$89,999
	1999	96	39	41%	$56,250
	1998	243	34	14%	$50,000
	1997	117	32	27%	$49,653
PROCESS & REACTION ENGINEERING	2005	221	23	10%	$91,658
	2004	194	26	13%	$83,617
	2003	117	26	22%	$87,564
	2002	72	30	42%	$75,600
	2001	101	27	27%	$73,914
	2000	137	32	23%	$64,892
	1999	84	28	33%	$64,601
	1998	51	21	41%	$70,198
	1997	69	21	30%	$67,347

CBET Funding Areas	Fiscal Year	Number of Proposals	Number of Awards	Funding Rate	Median Annual Size
SEPAR & PURIFICATION	2005	89	18	20%	$89,999
PROCESSES	2004	61	23	38%	$88,355
	2003	117	26	22%	$89,574
	2002	48	13	27%	$80,000
	2001	77	28	36%	$83,016
	2000	96	29	30%	$67,487
	1999	60	28	47%	$72,636
	1998	61	28	46%	$65,000
	1997	67	27	40%	$50,000
THERMAL TRANSPORT	2005	184	26	14%	$83,404
& THERM PROC	2004	170	29	17%	$83,559
	2003	112	30	27%	$87,185
	2002	70	24	34%	$84,030
	2001	67	21	31%	$73,683
	2000	83	18	22%	$73,542
	1999	93	30	32%	$84,118
	1998	135	27	20%	$63,267
	1997	106	35	33%	$60,054

6

Conclusions

The report summary and conclusions are provided below. Overall, the analysis was limited by a paucity of field-specific R&D funding and workforce data at the international level. Nonetheless, the members of the Panel have strong confidence in the conclusions provided in earlier sections and below.

The United States presently is and is expected to remain in the future among the world's leaders in all subareas of chemical engineering research with clear leadership in several.

The United States is currently among world leaders in all of the subareas of chemical engineering research and enjoys a leading position in both classical subareas as well as emerging areas including:

- transport processes;
- cellular and metabolic engineering;
- systems, computational, and synthetic biology;
- polymers;
- nanostructured materials;
- drug targeting and delivery systems;
- biomaterials;
- materials for cell and tissue engineering;
- fossil energy extraction and processing;
- air pollution;
- aerosol science and engineering;

- dynamics, control and operational optimization;
- safety and operability of chemical plants; and
- computational tools and information technology.

The United States is expected to maintain in the future its current position at the "Forefront" or "Among World Leaders" in all subareas of chemical engineering research. It is expected to expand and extend its current position in the following subareas:

- biocatalysis and protein engineering;
- cellular and metabolic engineering;
- systems, computational and synthetic biology;
- nanostructured materials;
- fossil energy extraction and processing;
- non-fossil energy;
- green engineering;

However, the strong U.S. position in transport processes; separations; heterogeneous catalysis; kinetics and reaction engineering; electrochemical processes; molecular and interfacial science and engineering; inorganic and ceramic materials; process development and design; and dynamics, control, and operational optimization has been weakened. Leadership in these core areas is now shared with Europe and in specific instances with Japan. Japan and other Asian countries are also particularly competitive in the materials-oriented research, e.g., polymers, inorganic and ceramic materials, biomaterials, and nanostructured materials. In addition, Europe is very competitive in the biorelated subareas of research while Japan is particularly strong in bioprocess engineering.

The Panel views the current research trends as healthy. At the same time, the group is concerned about the progressive erosion of U.S. positions in the core areas, because it is the strength in fundamentals that has enabled generations of chemical engineers to create new and highly competitive technologies for processes and products.

A strong manufacturing base, a strong culture and system of innovation, and the excellence and flexibility of the education and research enterprise have been and still are the major determinants of U.S. leadership in chemical engineering.

The keys to U.S. leadership in chemical engineering research have been the strength and global presence of the U.S. chemical, pharmaceutical, electronic, petroleum, biotechnology, and biomedical companies, the reach of the diverse U.S. economy, and the entrepreneurial ability of its researchers—entrepreneurial in both the academic and commercial sense. The rapid exploitation of new developments is facilitated by the extensive networks and collaborations among leading U.S. chemical engineering researchers that extend to all sectors of the U.S. economy and throughout the world. With the U.S. chemical companies well positioned to maintain and strengthen their global presence and reach, increasing numbers of new consumers, an essential prerequisite for the continued success of U.S. chemical engineering research is assured. However, as the chemical industry becomes progressively (a) more focused on new applications for existing chemical and material platforms rather than inventing big "blockbusters" like nylon and then engineering the lowest cost manufacturing approaches, and (b) more product-centric with greater emphasis on the market trends, it is the strength of the innovation system that will sustain and expand the competitiveness of U.S. research in chemical engineering: innovation in education; innovation in research directions with broad and deep impact; innovation in the modes of carrying out research in collaboration with other researchers, government, and industry; and innovation in the modes of technology transfer to large chemical companies or small startups. Federal programs that encourage research consortia and partnerships in the private sector and that fund precompetitive research at academic institutions, national laboratories, and small to medium-sized companies provide a strong impetus to the development of innovative technologies for chemicals, materials, products, and processes.

U.S. graduate education and research experience in chemical engineering has a high level of intellectual diversity, which intertwines with rich human diversity; chemical engineering in the United States has been the destination of choice of human talent from around the world. It has attracted young people with experimental, theoretical and computational, academic, industrial, policy-making, financial, or commercial bents. In addition, U.S. educational programs in chemical engineering have endowed chemical engineering researchers not only with important subject-matter knowledge, which makes them flexible, but also with a keen learning agility, making them quickly adaptable to new "hot topics" and more responsive to competitive pressures. Indeed, analysis of publications and patent data clearly demonstrates that U.S. chemical engineering researchers move much faster in defining or contributing to new areas of research than their counterparts throughout the world.

Moving faster allows one to establish leadership, but a flexible balance among all the key determinants is required to sustain leadership. These determinants include:

- availability of many options for funding research and entrepreneurial developments,

- creation of opportunities that enhance the diversity of the U.S. talent base,
- continuous improvements in research quality and productivity through greater unification of diverse new elements in the field and expansion in multidisciplinary collaborations, and
- balanced treatment of the short-term focus of the U.S. innovation system against the sustained health of the education and research enterprise that underpins the success of commercial innovation.

The agility and flexibility of the U.S. chemical engineering researcher is a major source of competitive advantage. The Panel believes that it needs to be preserved and strengthened. Therefore, federal programs and industrial support that encourage innovation in graduate chemical engineering education could have a deep and long-lasting impact on research competitiveness. Such programs may include support for initiatives leading to

- development of cohesive new core curricula naturally integrating physical, chemical, and biological phenomena at all spatial and temporal scales;
- enhanced interaction with researchers from other disciplines, particularly chemistry, biology, physics, materials science and engineering, and with industrial researchers;
- experience in defining, developing, and deploying innovation projects inspired from research results; and
- opportunities for international cooperation.

Shifting federal and industry funding priorities, a potential reduction in attracting human talent, domestic or foreign, and a narrowing of the discipline's technology breadth could diminish the United States' ability to turn today's scientific and technical discoveries into tomorrow's leading jobs in industry and education.

U.S. leadership in the various areas of chemical engineering is not assured for the future. In contrast to opportunities of leadership, there are current developments that could hurt the ability of the United States to capitalize on these opportunities. These include shifting funding priorities by federal agencies, reductions in industrial support of academic research in the United States in favor of academic support in other countries, potential decreases in the supply of talented foreign graduate students, reduced attractiveness of chemical engineering as a career path for the most talented U.S. citizens and permanent residents, shrinking of U.S.-based research laboratories by major chemical companies, and lack of attention to research into methods for shortening the development and implementation cycle for new chemicals, materials, processes, and products.

The Panel's analysis of the data has clearly indicated the weakening of the U.S. position in several areas of chemical engineering research, especially the core areas with a significant component of fundamental research. This weakening is presently confined in volume of output, but may eventually extend to include quality and impact. Stopping and reversing this erosion is of critical importance.

Human talent is at the heart of leadership. Attracting the best young people to the chemical engineering research enterprise is a prerequisite to sustaining our leadership. International competition for talent is heating up, and the winners will be determined by their ability to attract and retain human talent. The U.S. education system in chemical engineering has achieved excellence, which has been acknowledged throughout the world, and continues to attract top talent from other countries, especially those that lack adequate programs for training research leaders. There is concern that improvements in graduate programs in developing countries will not only meet their own needs for building indigenous research, but will attract home the top researchers and students who currently reside in the United States. To compound this potential threat, smaller numbers of talented young Americans choose science and engineering as their profession, leading to a smaller pool of talented individuals from which to draw the next generation of chemical engineering researchers. Starting salaries for Ph.D. chemical engineers, although still quite attractive in relation to salaries for Ph.D.s from sciences and other engineering disciplines, have just barely kept pace with inflation over the past 25 years. Today, other professions offer higher financial incentives and draw increasing fractions of talented young Americans. Similar trends have been observed in Europe, Japan, and in developing countries such as India.

The dynamic range of chemical engineering research over many spatial and temporal scales, across a broad range of products and processes, and throughout the vast variety of industries and social needs it serves, has been a profound force of innovation and competitiveness. This dynamic range is presently at risk. Federal funding opportunities are plentiful in support of research at "small" scales, molecular, nano, and biomedical. With the exception of ethanol plants in the midwestern United States, the U.S. chemical industry is choosing to build new plants not in the United States but in emerging economies. In addition, chemical industry support for academic research has been reduced and directed towards narrowly targeted developments with quick payback.

Newly established research centers and research consortia have enhanced the centrifugal forces of chemical engineering research towards the periphery of the field where they interact with various other disciplines.

Although such developments may be seen as results of a process that enhances research efficiency and effectiveness and directs resources towards high value-added research outcomes, they nevertheless erode the historical technology underpinnings of a successful national research enterprise and put the United States in a highly vulnerable position when "lost" or "deteriorated" competencies are once again needed for future technologies. An important example of this is the inevitable need for alternative energy sources. Virtually all of the options being explored today will rely heavily on traditional chemical engineering for implementation. If the United States becomes a nation of "nanomaterial-makers," the country may be first to exploit nanomaterials for new energy sources, but will lack the wherewithal to implement a total solution. At best this weakness will delay implementation; at worst the United States will need to "buy" technology from abroad and suffer the economic consequences. The Panel believes that addressing this issue is of critical importance for addressing national needs in energy and the environment and preserving U.S. competitiveness in the future.

Appendix A

Statement of Task

At the request of the National Science Foundation Engineering Directorate, the National Academies will perform an international benchmarking exercise to determine the standing of the U.S. research enterprise relative to its international peers in the fields of chemical engineering. The benchmarking exercise will address the following:

- What is the position of U.S. research in chemical engineering relative to that in other regions or countries?
- What are the key factors influencing relative U.S. performance in chemical engineering (i.e., human resources, equipment, infrastructure, etc.)?
- On the basis of current trends in the United States and worldwide, extrapolate to the U.S. relative position in the near and longer-term future.

Appendix B

Panel Biographies

CHAIR

George Stephanopoulos (NAE) is the Arthur D. Little Professor of Chemical Engineering at the Massachusetts Institute of Technology. His research interests include product and process development and design, process operations and control, and integrated computer-aided environments for process systems engineering. He was the Director of MIT-LISPE (Laboratory for Intelligent Systems in Process Engineering) and he has advised numerous chemical and engineering systems companies in the United States, Europe, and Japan. During the period 2000-2006 he served as Chief Technology Officer and Board Member of Mitsubishi Chemical Corporation, Japan. He received his Ph.D. from the University of Florida, ME from McMaster University, and a diploma of chemical engineering from the National Technical University of Athens.

MEMBERS

Pierre Avenas is delegate for research in ParisTech (Paris Institute of Technology), an association which brings together 11 French engineering universities located in or near Paris. He is the former head of research and development for Atofina Chemicals. He is also a member of the IDEA League working group on research. IDEA League is a strategic alliance between Imperial College London, TU Delft, ETH Zürich, RWTH Aachen, and ParisTech.

William F. Banholzer (NAE) is Corporate Vice President and Chief Technology Officer of The Dow Chemical Company, located in Midland, Michigan. He is a member of the Office of the Chief Executive (OCE) and leads Dow's research and development activities across the globe. Banholzer joined Dow in July 2005 from General Electric Company, where he was Vice President of Global Technology at GE Advanced Materials, responsible for worldwide technology and engineering. Banholzer holds a bachelor's degree in chemistry from Marquette University and earned master's and doctorate degrees in chemical engineering from the University of Illinois.

Gary S. Calabrese is Vice President and Chief Technology Officer of Rohm & Haas Company, responsible for a 2000+ member global technical organization of scientists, engineers and technicians with over 30 worldwide locations, including a new research center in China. Prior to joining Rohm & Haas, Dr. Calabrese began his industrial career at Polaroid Corporation in 1983 as a research chemist. In 1989 he joined the Shipley Company to work as a group leader in new product development. In 1992 Shipley became a wholly-owned subsidiary of Rohm and Haas, and in 1994 Dr. Calabrese was named Shipley's North American Director of Engineering, responsible for scaling up manufacturing processes for new products, customer technical support, and plant engineering. He returned to product development in 1997 as Global Director of R&D for the Microelectronics Materials business, and was named Vice President and Chief Technology Officer for what is now known as Rohm & Haas Electronic Materials two years later. In this position he was responsible for a global technology organization with more than 300 members in seven locations including Japan and Korea. Dr. Calabrese earned his BS in chemistry from Lehigh University, and his Ph.D. in inorganic chemistry from the Massachusetts Institute of Technology

Douglas S. Clark is Professor of Chemical Engineering at the University of California at Berkeley. His research is in the field of biochemical engineering, with particular emphasis on enzyme technology and bioactive materials, extremophiles and extremophilic enzymes, cell culture, and metabolic flux analysis. He received a B.S. from the University of Vermont and a Ph.D. from the California Institute of Technology.

L. Louis Hegedus (NAE) retired in 2006, after 10 years of service, as the Senior Vice President of Research and Development for Arkema, Inc., a diversified chemical company headquartered in Paris. He was responsible for all R&D in North America and R&D coordination between the United States and France. His previous career positions include 16 years with W. R. Grace, where he was a Research Vice President for Specialty Chemicals, and 8 years with the General Motors Research Laboratories, where he contributed to the development of the catalytic converter for automobile emission control. He received his Ph.D. in chemical engineering from the University of California, Berkeley and an MS in chemical engineering from the Technical University of Budapest. Dr. Hegedus is a past Chairman of the chemical engineering section of the NAE, and a past Chairman of the Council for Chemical Research.

Eric W. Kaler is the Dean of the College of Engineering and the Elizabeth Inez Kelley Professor of Chemical Engineering at the University of Delaware. Dr. Kaler's research focuses on colloid and surfactant science and engineering, complex fluid dynamics, materials synthesis, and small angle scattering. He also has an interest in polymer science, and has worked on all varieties of surfactant materials and structures, including emulsions, microemulsions, micelles, vesicles, and liposomes. He received his Ph.D. from the University of Minnesota and his bachelor's from the California Institute of Technology.

Julio M. Ottino (NAE) is currently Dean of the Robert R. McCormick School of Engineering and Applied Sciences at Northwestern University and holds the titles of Distinguished Robert R. McCormick Institute Professor and Walter P. Murphy Professor of Chemical and Biological Engineering. He was Chairman of the department of Chemical Engineering during 1992-2000. Ottino's research has impacted fields as diverse as fluid dynamics, granular dynamics, microfluidics, geophysical sciences, and nonlinear dynamics and chaos and has appeared on the covers of and *Nature, Science, Scientific American,* the *Proceedings of the National Academy of Sciences of the US.* He is currently a senior advisor to Unilever, was a member of the Technical Board of Dow Chemical, and was a member of the 2004 EPSRC/Royal Academy International Review of Engineering in the United Kingdom. Dr. Ottino received the Alpha Chi Sigma Award and William H. Walker of the American Institute of Chemical Engineers and was the Danckwerts Lecturer in London. He has been a Guggenheim Fellow and a Sigma Xi Lecturer, and is a Fellow of the American Physical Society, a Member of the National Academy of Engineering and the American Academy of Arts and Sciences.

Nicholas A. Peppas (NAE) is the Fletcher S. Pratt Chair in Engineering at the University of Texas at Austin with appointments in Chemical Engineering, Biomedical Engineering, and Pharmacy. He is also the Director of the Center on Biomaterials, Drug Delivery and Bionanotechnology. He has collaborated with numerous international companies in the polymers, pharmaceutical, and medical fields and has been a visiting professor at the universities of Paris, Berlin, Geneva, Parma, Naples, Pavia, Athens, Hacettepe (Ankara), Hebrew (Jerusalem), Hoshi (Tokyo) and Nanyang (Singapore). He is the cofounder of several biotechnology companies. He received a diploma of engineering from the National Technical University of Athens, Greece and a ScD from the Massachusetts Institute of Technology, both in chemical engineering.

John D. Perkins is Vice President and Dean of the Faculty of Engineering & Physical Sciences at the University of Manchester. Professor Perkins was until recently Principal of the Faculty of Engineering and Courtaulds Professor of Chemical Engineering at Imperial College London. His academic career spans periods at Cambridge University and at the University of Sydney as well as Imperial College. He has industrial experience with Shell and with ICI, in the United Kingdom and in Australia, and has acted as a consultant in process control and process modeling and simulation for a number of companies around the world. He is the author of a number of authoritative reports and has managed several industrial consortia projects aimed at benchmarking and introducing advanced control methods into industrial practice.

Julia M. Phillips (NAE) is the Director of the Physical, Chemical, and Nano Science Center at Sandia National Laboratories. Phillips began her career at Sandia in 1995 after 14 years at AT&T Bell Laboratories. She has a PhD in applied physics from Yale University and a BS in physics from the College of William and Mary. Her research has been in the areas of epitaxial metallic and insulating films on semiconductors, high-temperature superconducting, ferroelectric and magnetic oxide thin films, and novel transparent conducting materials.

Adel F. Sarofim (NAE) is Presidential Professor in the College of Engineering, University of Utah and Senior Technical Advisor to Reaction Engineering International in Salt Lake City. Dr. Sarofim is the author and coauthor of over 200 papers on the subjects of radiative heat transfer, furnace design, circulation patterns in glass melts, the freeze process for desalination, nitric oxide formation in combustion systems, combustion-generated aerosols, soot and polycyclic aromatic hydrocarbon formation, and the characterization of carbon structure and reactivity. He received a BA in chemistry from Oxford University and an SM and ScD in chemical engineering from the Massachusetts Institute of Technology.

Jackie Y. Ying was born in Taipei, and raised in Singapore and New York. She received her BE from The Cooper Union and PhD from Princeton University. She was an NSF-NATO Postdoctoral Fellow and Alexander von

Humboldt Research Fellow at the Institute for New Materials, Germany. She joined the Chemical Engineering faculty of Massachusetts Institute of Technology in 1992, and was promoted to the rank of Professor in 2001. She has been the Executive Director of the Institute of Bioengineering and Nanotechnology in Singapore since 2003. Ying has been recognized with a number of awards for her research in nanostructured materials, including the American Ceramic Society Ross C. Purdy Award, David and Lucile Packard Fellowship, Office of Naval Research and National Science Foundation Young Investigator Awards, Camille Dreyfus Teacher-Scholar Award, Royal Academy of Engineering ICI Faculty Fellowship, American Chemical Society Faculty Fellowship Award in Solid-State Chemistry, Technology Review TR100 Young Innovator Award, American Institute of Chemical Engineers Allan P. Colburn Award. She was elected a World Economic Forum Global Young Leader, and a member of the German Academy of Natural Scientists, Leopoldina.